Salivary Gland Neoplasms

Advances in Oto-Rhino-Laryngology

Vol. 78

Series Editor

Patrick J. Bradley Nottingham

Salivary Gland Neoplasms

Volume Editors

Patrick J. Bradley Nottingham
David W. Eisele Baltimore, Md.

49 figures, 33 in color, and 25 tables, 2016

Basel · Freiburg · Paris · London · New York · Chennai · New Delhi ·
Bangkok · Beijing · Shanghai · Tokyo · Kuala Lumpur · Singapore · Sydney

Patrick J. Bradley, MBA, FRCSEd, FRCS (Hon), FRACS (Hon), FRCSLT (Hon)
Emeritus Honorary Professor, Department ORL-HNS
Nottingham University Hospitals
Queens Medical Centre Campus, Derby Road
Nottingham NG7 2UH (UK)

David W. Eisele, MD, FACS
Andelot Professor and Director
Department of Otolaryngology – Head and Neck Surgery
Johns Hopkins University School of Medicine
601 N. Caroline St., Suite 6210
Baltimore, MD 21287-0910 (USA)

Library of Congress Cataloging-in-Publication Data

Names: Bradley, Patrick J., 1949- , editor. | Eisele, David W., editor.
Title: Salivary gland neoplasms / volume editors, Patrick J. Bradley, David W. Eisele.
Other titles: Advances in oto-rhino-laryngology ; v. 78.
Description: Basel ; New York : Karger, [2016] | Series: Advances in oto-rhino-laryngology ; vol. 78 | Includes bibliographical references and index.
Identifiers: LCCN 2016002929| ISBN 9783318058017 (hard cover : alk. paper) | ISBN 9783318058024 (electronic version)
Subjects: | MESH: Salivary Gland Neoplasms--diagnosis | Salivary Gland Neoplasms--pathology | Salivary Gland Neoplasms--therapy
Classification: LCC RC280.S3 | NLM WI 230 | DDC 616.99/4316--dc23 LC record available at http://lccn.loc.gov/2016002929

Bibliographic Indices. This publication is listed in bibliographic services, including Current Contents® and Index Medicus.

Disclaimer. The statements, opinions and data contained in this publication are solely those of the individual authors and contributors and not of the publisher and the editor(s). The appearance of advertisements in the book is not a warranty, endorsement, or approval of the products or services advertised or of their effectiveness, quality or safety. The publisher and the editor(s) disclaim responsibility for any injury to persons or property resulting from any ideas, methods, instructions or products referred to in the content or advertisements.

Drug Dosage. The authors and the publisher have exerted every effort to ensure that drug selection and dosage set forth in this text are in accord with current recommendations and practice at the time of publication. However, in view of ongoing research, changes in government regulations, and the constant flow of information relating to drug therapy and drug reactions, the reader is urged to check the package insert for each drug for any change in indications and dosage and for added warnings and precautions. This is particularly important when the recommended agent is a new and/or infrequently employed drug.

All rights reserved. No part of this publication may be translated into other languages, reproduced or utilized in any form or by any means electronic or mechanical, including photocopying, recording, microcopying, or by any information storage and retrieval system, without permission in writing from the publisher.

© Copyright 2016 by S. Karger AG, P.O. Box, CH–4009 Basel (Switzerland)
www.karger.com
Printed in Germany on acid-free and non-aging paper (ISO 9706) by Kraft Druck GmbH, Ettlingen
ISSN 0065–3071
e-ISSN 1662–2847
ISBN 978–3–318–05801–7
e-ISBN 978–3–318–05802–4

Contents

VII Preface
Bradley, P.J. (Nottingham); Eisele, D.W. (Baltimore, Md.)

1 Classification of Salivary Gland Neoplasms
Bradley, P.J. (Nottingham)

9 Frequency and Histopathology by Site, Major Pathologies, Symptoms and Signs of Salivary Gland Neoplasms
Bradley, P.J. (Nottingham)

17 Molecular Pathology and Biomarkers
Ha, P.K (San Francicso, Calif.); Stenman, G. (Gothenburg)

25 Evaluation: Imaging Studies
Kontzialis, M. (Baltimore, Md.); Glastonbury, C.M. (San Francisco, Calif.); Aygun, N. (Baltimore, Md.)

39 Evaluation: Fine Needle Aspiration Cytology, Ultrasound-Guided Core Biopsy and Open Biopsy Techniques
Howlett, D.C. (Eastbourne); Triantafyllou, A. (Liverpool)

46 Facial Nerve Monitoring
Guntinas-Lichius, O. (Jena); Eisele, D.W. (Baltimore, Md.)

53 Surgery for Benign Salivary Neoplasms
Gillespie, M.B. (Charleston, S.C.); Iro, H. (Erlangen)

63 Recurrent Benign Salivary Gland Neoplasms
Witt, R.L. (Newark, Del.); Nicolai, P. (Brescia)

71 Prognostic Scoring for Malignant Salivary Gland Neoplasms
Vander Poorten V. (Leuven); Guntinas-Lichius, O. (Jena)

83 Surgery for Primary Malignant Parotid Neoplasms
Deschler D.G. (Boston, Mass.); Eisele, D.W. (Baltimore, Md.)

95 Metastatic Cancer to the Parotid
Clark, J. (Sydney, N.S.W.); Wang, S. (San Francisco, Calif.)

104 Surgery for Malignant Submandibular Gland Neoplasms
Silver, N.L.; Chinn, S.B. (Houston, Tex.); Bradley, P.J. (Nottingham); Weber, R.S (Houston, Tex.)

113 Surgery for Malignant Sublingual and Minor Salivary Gland Neoplasms
Bradley, P.J. (Nottingham); Ferris, R.L. (Pittsburgh, Pa.)

- 120 **Facial Reconstruction and Rehabilitation**
  Guntinas-Lichius, O. (Jena); Genther, D.J.; Byrne, P.J. (Baltimore, Md.)
- 132 **Management of Regional Metastases of Malignant Salivary Gland Neoplasms**
  Medina, J. (Oklahoma City, Okla.); Zbären, P. (Bern); Bradley, P.J. (Nottingham)
- 141 **Indications for Salivary Gland Radiotherapy**
  Thomson, D.J.; Slevin, N.J. (Manchester); Mendenhall, W.M. (Gainesville, Fla.)
- 148 **Chemotherapy and Targeted Therapy**
  Sen, M.; Prestwich, R. (Leeds)
- 157 **Management of Inoperable Malignant Neoplasms**
  Kiess, A.P.; Quon, H. (Baltimore, Md.)
- 168 **Management of Recurrent Malignant Salivary Neoplasms**
  Merdad, M.; Richmon, J.D.; Quon, H. (Baltimore, Md.)
- 175 **Salivary Gland Neoplasms in Children and Adolescents**
  Bradley, P.J. (Nottingham); Eisele, D.W. (Baltimore, Md.)
- 182 **Distant Metastases and Palliative Care**
  Glazer, T.A.; Shuman, A.G. (Ann Arbor, Mich.)
- 189 **Quality of Life after Salivary Gland Surgery**
  Wax, M.K. (Portland, Oreg.); Talmi, Y.P. (Tel Hashomer)
- 198 **Salivary Gland Neoplasms: Future Perspectives**
  Eisele, D.W. (Baltimore, Md.); Bradley, P.J. (Nottingham)

- 200 **Author Index**
- 201 **Subject Index**

Preface

Salivary gland neoplasms are uncommon, and for the majority of clinicians, exposure to their diagnosis and management is gained over the clinician's lifetime. In 1991, the World Health Organization published that there were 12 recognised benign and 19 malignant primary epithelial salivary gland neoplasms, and a decade later, in 2001, it published an update stating that there were now 13 benign and 24 malignant primary epithelial salivary gland neoplasms. Since that time, a number of 'new' primarily malignant primary epithelial salivary gland neoplasms have been described.

This volume on salivary gland neoplasms is considered by the volume editors to be a beneficial educational resource for clinicians involved in the management of patients with salivary gland disorders, especially when considering the possible diagnosis of a salivary gland neoplasm. This volume is composed of twenty-three chapters, each written by world-renowned experts on salivary gland neoplasms. The intent for each chapter was to be concise, with concentration on the advances in this clinical area that have recently been made, while still summarising what is known. This editorial process has minimised duplication of information, except where it has been considered important, and each chapter has a bibliography of relevant publications that might be consulted should additional information be required.

The chapters cover the breadth of salivary gland neoplasms, from the major to the minor salivary glands, from benign to malignant, from primary to secondary, from children to adults, and from recurrent benign to malignant disease, and describe management of local regional metastasis, reconstructive procedures following radical parotidectomy, indications for radiotherapy, usage of chemotherapy and targeted therapy, management of distant metastasis and palliative care, quality of life and methods of prognostic scoring for malignant salivary gland neoplasms. The final chapter presents the volume editors' comments on the future perspective necessary to improve treatment outcomes and patients' quality of life.

Patrick J. Bradley, MBA, FRCS,
Nottingham, UK
David W. Eisele, MD, FACS,
Baltimore, Md., USA

Classification of Salivary Gland Neoplasms

Patrick J. Bradley

School of Medicine, The University of Nottingham, Nottingham University Hospitals, Queens Medical Centre Campus, Nottingham, UK

Abstract

Presently, there is no universal 'working' classification system acceptable to all clinicians involved in the diagnosis and management of patients with salivary gland neoplasms. The most recent World Health Organization Classification of Tumours: Head and Neck Tumours (Salivary Glands) (2005) for benign and malignant neoplasms represents the consensus of current knowledge and is considered the standard pathological classification based on which series should be reported. The TNM classification of salivary gland malignancies has stood the test of time, and using the stage groupings remains the current standard for reporting treated patients' outcomes. Many developments in molecular and genetic methods in the meantime have identified a number of new entities, and new findings for several of the well-established salivary malignancies need to be considered for inclusion in any new classification system. All clinicians involved in the diagnosis, assessment and treatment of patients with salivary gland neoplasms must understand and respect the need for the various classification systems, enabling them to work within a multidisciplinary clinical team environment.

© 2016 S. Karger AG, Basel

Introduction

The global annual incidence of all salivary gland neoplasm (SGNs) varies from 0.4 to 13.5 cases per 100,000 population, and the frequency of malignant salivary neoplasms ranges from 0.4 to 2.6 cases per 100,000 population [1]. Malignant epithelial SGNs are relatively uncommon and account for approximately 3–6% of all head and neck malignancies. SGNs have a special status in human neoplasia, as they have the most complex histopathology of any organ/tissue, with an exceptional breadth of different tumour types that exhibit a wide variety of microscopic appearances, often within a single lesion. Benign SGNs are estimated to present in clinical practice 5–7 times more frequently than their malignant counterparts, and the majority (>80%) are diagnosed as pleomorphic adenoma.

Table 1. WHO (2005) histological classification of epithelial tumours of the salivary glands [1]

Malignant epithelial (n = 24)	
Acinic cell carcinoma	Oncocytic carcinoma
Mucoepidermoid carcinoma	Salivary duct carcinoma
Adenoid cystic carcinoma	Adenocarcinoma (NOS)
Polymorphous (LG) adenocarcinoma	Myoepithelial carcinoma
Epithelial-myoepithelial carcinoma	Carcinoma ex-pleomorphic adenoma
Clear cell carcinoma (NOS)	Carcinosarcoma
Basal cell adenocarcinoma	Metastasising pleomorphic adenoma
Sebaceous carcinoma	Squamous cell carcinoma
Sebaceous lymphadenocarcinoma	Small cell carcinoma
Cystadenocarcinoma	Large cell carcinoma
LG cribriform cystadenocarcinoma	Lymphoepithelial carcinoma
Mucinous adenocarcinoma	Sialoblastoma
Benign epithelial (n = 10)	
Pleomorphic adenoma	Lymphadenoma
Myoepithelioma	Sebaceous
Basal cell adenoma	Non-sebaceous
Warthin tumour	Ductal papilloma
Oncocytoma	Inverted ductal papilloma
Canalicular adenoma	Intraductal papilloma
Sebaceous adenoma	Sialadenoma papilliferum
Cystadenoma	

LG = Low-grade; NOS = not otherwise specified.

History of Pathological Classification

Over the past 50 years, several pathological classification schemes have been proposed, and the Armed Forces Institute of Pathology and the World Health Organization (WHO) have been at the forefront of producing acceptable clinical and pathological systems. Salivary gland tumours are classified as benign and malignant epithelial neoplasms, mesenchymal (soft tissue) benign and malignant neoplasms, haematolymphoid tumours [Hodgkin lymphoma (HL), non-HL, low-grade B-cell lymphoma arising in mucosa-associated lymphoid tissue and extramedullary plasmacytoma] and secondary tumours [1].

Progress in histopathology is evident based on the identification and classification of 16 salivary gland epithelial tumour entities in the 1954 Armed Forces Institute of Pathology fascicle, compared to 36 entities in the 1996 publication. Between 1954 and 2005, the date of the most recent WHO publication, there has been a doubling in tumour categories for salivary gland epithelial tumours of both benign (from 6 to 10) and malignant (from 10 to 24) nature [1] (table 1). It has been recommended that when reporting future epidemiological surveys, the latest WHO classification should be used to characterise the data presented, even in small geographical regions [2].

Pure mesenchymal neoplasms are very uncommon when compared to epithelial SGNs and are reported to account for 19.9–4.7% of SGNs, with benign soft tissue lesions being more common than sarcomas. The ratio of benign to malignant mesenchymal tumours varies widely, ranging from 18:1 to 2.4:1 in the few reported

Table 2. Histological classification of mesenchymal tumours of the salivary gland (modified from Eveson et al. [1] and Cockarill et al. [4])

Benign mesenchymal	Malignant sarcoma
Vascular	
Haemangioma	Rhabdomyosarcoma
Lymphangioma	Haemangiopericytoma
Neural	
Neurofibroma (Schwannoma)	Liposarcoma
	Malignant fibrous histiocytosis
	Synovial sarcoma
Fibroblastic/myofibroblastic	
Nodular fasciitis	Malignant peripheral nerve sheath tumour
	Alveolar rhabdomyosarcoma
Fibroblastic/myofibroblastic	
Nodular fasciitis	Leiomyosarcoma
Myofibromatosis	Kaposi sarcoma
Fibroma	Langerhans cell histiocytosis
Solitary fibrous tumour	Endodermal sinus tumour
	Osteosarcoma
Lipoma	
Pleomorphic variety	Atypical lipomatous tumour
	Desmoplastic small round cell tumour
Granular cell tumour	Fibrosarcoma
Angiomyoma	Interdigitating dendritic cell tumour
Glomangioma	Myelosarcoma
Myxoma	Sarcoma ex-pleomorphic adenoma
Fibrous histiocytosis	Chondrosarcoma

series [3, 4]. Over 85% of soft tissue neoplasms arise in the parotid gland, and >10% involve the submandibular gland, whereas a tumour rarely arises in the sublingual gland [3]. Vascular neoplasms are the most common benign mesenchymal lesion, accounting for >40%, with 75–80% being haemangiomas appearing in the first decade of life. Most of the other vascular lesions are lymphangiomas (table 2). It is often difficult to determine whether a reported mesenchymal tumour arose from the salivary gland parenchyma or from the surrounding paraglandular connective tissue, and hence, these tumours have been referred to as arising in the 'parotid' or 'submandibular' region. A recent review lists some 42 histological entities (top 20 listed in table 2) that have been reported to affect the major salivary glands [4].

Involvement of the salivary glands in HL is exceedingly rare, though cases of classical HL and nodular lymphoblastic predominant HL have been reported. It is difficult to determine whether such cases are they are truly primary HL or part of a disseminated disease process. Most non-HLs are B-cell lymphomas and may be primary or secondary. The parotid gland is most affected (75%), followed by the submandibular gland (20%). Most patients are over 60 years of age and present with a palpable mass, with pain and tenderness being uncommon [5].

Table 3. Risk stratification of WHO (2005)-recognised epithelial salivary gland malignancies (modified from Seethala [12])

Low risk	High risk
Acinic cell carcinoma	Sebaceous carcinoma and lymphadenocarcinoma
LG mucoepidermoid carcinoma*	HG mucoepidermoid carcinoma*
Epithelial-myoepithelial carcinoma	Adenoid cystic carcinoma†
Polymorphous LG adenocarcinoma	Mucinous adenocarcinoma
Clear cell carcinoma	Squamous cell carcinoma
Basal cell adenocarcinoma	Small cell carcinoma
LG salivary duct carcinoma	Large cell carcinoma
(LG cribriform cystadenocarcinoma)	Salivary duct carcinoma
Myoepithelial carcinoma	Lymphoepithelial carcinoma
Oncocytic carcinoma	Metastasising pleomorphic adenoma
Carcinoma ex-pleomorphic adenoma	Carcinoma ex-pleomorphic adenoma (widely
(intracapsular/minimally invasive or LG histology)	invasive or HG histology)
Sialoblastoma	Carcinosarcoma
Adenocarcinoma NOS (LG)	Adenocarcinoma NOS (HG)
Cystadenocarcinoma (LG)*	Cystadenocarcinoma (HG)*

* Intermediate-grade variants of these tumours are controversial regarding the assignment of risk. For mucoepidermoid carcinoma, this may depend on the grading scheme used. For adenocarcinoma NOS, there is little data, but what is present suggests that intermediate-grade variants should be placed in the high-risk group.
† Adenoid cystic carcinomas are all considered high risk in terms of local recurrence, but only solid adenoid cystic carcinoma (i.e. high-grade pattern) is considered high risk for nodal metastasis. NOS = Not otherwise specified; LG = low-grade; HG = high-grade.

It has been argued that any classification scheme should be considered dynamic, continuing to evolve as new light is shed on the biological behaviour of both established and newly described entities by clinical, pathological, molecular and genetic studies, resulting in frequent revisions of the 'old classification systems' via an evolutionary process. At best, a current classification system represents no more than the consensus of those individuals involved at the time of its formulation. Any tendency to regard a classification as immutable should be firmly resisted [6].

Clinician Stakeholders

It is of importance that the surgeon and the surgical pathologist have professional respect for each other's approaches to their working SGN classification and method of neoplasm diagnosis. To the surgeon, some tumour classifications and taxonomy appear to be rather esoteric, serving as an academic exercise of limited practical value within the clinical situation (table 3).

In the past, the pathological analysis of salivary gland tumours focused on relating histological tumour types to their cells of origin (histogenesis), given the presence of stem cell progenitors in the proximal and distal regions of the duct system. This would explain why salivary duct carcinoma (SDC), mucoepidermoid carcinoma (MEC) and adenocarcinoma are derived from the main duct (high-grade) and why tumours containing epithelial and myoepithelial cells develop from the terminal (intercalated) duct (low-grade). A more recently proposed theory (morphogenesis) relates to cell differentiation derived from differential gene expression by a stem cell in conjunction with tumour matrix production, rather than to a specific proposed cell of origin. Based on these two theories, two types of classification schemes for SGNs are possible. The taxonomic classification focuses on the inter-relationships of various subtypes of SGNs. The other classification is more of a 'working classification' in which the neoplasms are classified based on their unique histological features and biological behaviour [6]. The surgical pathologist's primary function is to produce as precise a diagnosis as possible. It is only by evaluation of subsequent clinical experience with SGNs' behaviour and responses to a variety of treatment protocols that the relative importance of an individual diagnosis can be established. It is in this area of precise diagnosis that the value of an extended classification of salivary gland tumours becomes most evident. Broad prognostic groupings and tumour grading are of no help if the initial diagnostic category is not accurate [6].

Histotyping Classification

There is frequently no uniform relationship between histotype and biological behaviour [7, 8]. Only population-based studies are a true reflection of the distribution of the 24 histotypes of malignant SGNs. In population-based studies, the majority of major SGNs are acinic cell carcinoma (AcCC; 15–17%), adenoid cystic carcinoma (ACC; 16–27%) or MEC (14.5–19%). In most studies of minor salivary glands, ACC and MEC account for up to 89% of histological types, and ACC (32–71%) and MEC (15–38%) far outnumber adenocarcinoma (not otherwise specified), ACC and polymorphous low-grade adenocarcinoma, epithelial myoepithelial-myoepithelial carcinoma, and carcinoma ex-pleomorphic adenoma [6]. Histotyping of a salivary gland carcinoma is challenging and is marked by high reclassification rates following slide review of historical series [9, 10]. Substantial interobserver variability between pathologists also exists. The many immunohistochemical stains currently available, however, have helped to improve the accuracy of tumour classification. Reclassification, inter-observer variability, geographical variation and referral bias all contribute to disparities in the published histological distribution [9].

Histopathological Grading Classification

The pathologist is also expected to make some comment on the grade of malignant salivary gland tumours (high-, intermediate- or low-grade), as the grade is supposed to reflect the inherent biological nature of the tumour (aggressive, intermediate or indolent) [9, 11, 12]. Some tumour families are known to be high grade and are thus considered to be of high risk or to have biological aggressiveness (most high-grade tumours: MEC, SDC, and squamous cell carcinoma), although some are low grade, such as AcCC, polymorphous low-grade adenocarcinoma, and adenocarcinoma (not otherwise specified). In the past, grading methods have shown poor inter- and intra-examiner consistency, with a low independent prognostic power in multivariate analysis, because grading parallels other important prognostic factors that are more reliably reproducible, such as age, stage (size and nodal status), and the presence of perineural growth [12, 13]. However, histological grade and thus, indirectly, tumour type rank highly as important prognosticators and should continue to be reported [14]. High-grade salivary carcinomas have a 5-year survival of ≈40%, whereas low- and intermediate-grade tumours have a 5-year survival of 85–90%. In large patient series, histological grade was an independent predictor of outcome in multivariate analysis [12]. Table 3 divides all of the described malignant epithelial SGNs listed in the WHO classification (2005) scheme into low-risk and high-risk categories based on a combination of entities defined by their behaviour and grade [12].

Table 4. Primary tumour (T) major salivary glands (parotid, submandibular and sublingual glands)

TX	Primary tumour cannot be assessed
T0	No evidence of primary tumour
T1	Tumour 2 cm or less in greatest dimension, without extra-parenchymal extension*
T2	Tumour more than 2 cm but not more than 4 cm in greatest dimension, without extra-parenchymal extension*
T3	Tumour more than 4 cm and/or tumour with extra-parenchymal extension*
T4a	Tumour invades skin, mandible, ear canal, or facial nerve
T4b	Tumour invades base of skull or pterygoid plates or encases carotid artery

* Extra-parenchymal extension is clinical or microscopic evidence of invasion of soft tissue or nerves, except in the cases listed under T4a and T4b. Microscopic evidence alone does not indicate extra-parenchymal extension for classification purposes.

Table 5. Regional lymph nodes (N)

NX	Regional lymph nodes cannot be assessed
N0	No regional lymph node metastasis
N1	Metastasis in a single ipsilateral lymph node, 3 cm or less in greatest dimension
N2	Metastasis, as specified in N2a, N2b, and N2c below
N2a	Metastasis in a single ipsilateral lymph node, more than 3 cm but not more than 6 cm in greatest dimension
N2b	Metastasis in multiple ipsilateral lymph nodes, none more than 6 cm in greatest dimension
N2c	Metastasis in bilateral or contralateral lymph nodes, none more than 6 cm in greatest dimension
N3	Metastasis in a lymph node, more than 6 cm in greatest dimension

Molecular Biology and Genetic Classification

Molecular and genetic markers are being investigated for their potential role and may prove more reliable in predicting outcomes. Molecular targets such as stem cell factor receptor, the cell cycle oncogene nuclear factor-κB in and vascular endothelial growth factor in ACC and human epidermal growth factor receptor 2 in all types of salivary malignancies raised initial hopes of a significant breakthrough in the addition of molecular targeted therapy [13]. To date, the results of systemic treatment by means of cytotoxic chemotherapy or targeted molecular therapies remain disappointing. More recently, the identification of specific translocations may revolutionise the way that salivary tumours are considered and may have a major impact on future diagnostic practice [7, 8, 15].

Since the WHO classification of 2005, there have been several important developments in the pathological analysis of salivary gland malignancies, and particularly molecular methods [15]. Two new and not uncommon carcinomas have been described, namely, mammary analogue secretory carcinoma and cribriform adenocarcinoma of the tongue and minor salivary glands, in addition to new findings in several well-established entities: MEC, ACC, epithelial-myoepithelial carcinoma, hyalinising clear cell carcinoma, and SDC. As a result, histopathologists have outlined their 'wish list' of revised and new entities of malignant SGNs to be included in any anticipated new version of the WHO malignant epithelial salivary gland classification [16].

TNM Stage Classification

All malignant tumours are staged according to the TNM classification system [17] (tables 3–5). The TNM system describes the anatomical extent of disease based on the assessment of three components: T – the extent of the primary tumour, N – the absence or presence and extent of regional nodal metastases, and M – the absence or presence of distant metastases. For malignant salivary tumours located in minor salivary glands, the TNM system used is the same as that used for squamous cell carcinoma at a similar anatomical site in the head and neck. Stages I/II are considered to be early disease, and Stages II/IV are late or advanced disease. The salivary gland disease stage recorded is a reliable prognostic indicator of patient survival; for advanced disease, the prognosis is worse.

Table 6. Stage grouping

Stage I	T1N0M0
Stage II	T2N0M0
Stage III	T3N0M0
	T1, T2, T3, N1, M0
Stage IVA	T1, T2, T3, N2, M0
	T4a, N0, N1, N2, M0
Stage IVB	T4b, Any N, M0
	Any T, N3, M0
Stage IVC	Any T, Any N, M1

Conclusion

Presently, there is no universal 'working' classification system acceptable to all clinicians involved in the diagnosis and management of patients with SGNs.

References

1 Eveson JW, Auclair P, Gnepp DR, et al: Tumours of the salivary glands; in Barnes L, Eveson JW, Reichart P, et al (eds): World Health Organisation Classification of Tumours. Pathology and Genetics of Head and Neck Tumours. Lyon, IARC Press, 2005, pp 212–215.
2 Bello IO, Salo T, Dayan D, et al: Epithelial salivary gland tumors in two distant geographical locations, Finland (Helsinki and Oulo) and Israel (Tel Aviv). Head Neck Pathol 2012;6:224–231.
3 Gnepp DR: Soft tissue tumors; in Barnes L, Eveson JW, Reichart P, et al (eds): World Health Organisation Classification of Tumours. Pathology and Genetics of Head and Neck Tumours. Lyon, IARC Press, 2005, p 275.
4 Cockarill CC, Daram S, El-Naggar AK, et al: Primary sarcomas of the salivary glands; case series and literature review. Head Neck 2013;35:1551–1117.
5 Chan AC, Chan JK, Abbondanzo SL: Haemato-lymhoid tumours; in Barnes L, Eveson J W, Reichart P, et al (eds): World Health Organisation Classification of Tumours. Pathology and Genetics of Head and Neck Tumours. Lyon, IARC Press, 2005, pp 277–280.
6 Eveson J: The WHO histological classification of salivary gland tumours; is it over elaborate for clinical use?; in McGurk M, Renahan A (eds): Controversies in the Management of Salivary Gland Disease, ed 1. Oxford, Oxford University Press, 2001, pp 25–29.

7 Bell D, Hanna EY: Salivary gland cancer: biology and molecular targets for therapy. Curr Oncol Rep 2012;14:166–174.
8 Leivo I: Insights into a complex group of neoplastic disease: advances in histopathologic classification and molecular pathologic classification. Acta Oncol 2006;45:662–668.
9 Vander Poorten V, Meulemans J, Delaere P, et al: Molecular markers and chemotherapy for advanced salivary cancer. Curr Otorhinolaryngol Rep 2014;2:85–96.
10 Bjørndal K, Krogdahl A, Therkildsen MA, et al: Salivary gland carcinoma in Denmark 1990–2005. a national study of incidence, site, and histology. Results of the Danish Head and Neck Cancer Group (DHANCA). Oral Oncol 2011;47: 677–682.
11 Bansal AK, Bindal R, Kapoor C, et al: Current concepts in diagnosis of unusual salivary gland tumours. Dental Res J 2012;9(suppl 1):S9–S18.
12 Seethala RR: An update on grading of salivary gland carcinomas. Head Neck Pathol 2009;3:69–77.
13 Vander Poorten V, Hunt J, Bradley PJ, et al: Diagnosis and management of parotid carcinoma with special focus on recent advances in molecular biology. Head Neck 2012:34:429–440.
14 Walvekar RR, Andrade Filho PA, Seethala RR, et al: Clinicopathologic features as stronger prognostic factors than histology or grade in risk stratification of primary parotid malignancies. Head Neck 2011;33:225–231.
15 Simpson RHW, Skalova A, Di Palma S, et al: Recent advances in the diagnostic pathology of salivary carcinomas. Virchows Arch 2014;465:371–384.
16 Proceedings of the North American Society of Head and Neck Pathology Companion Meeting, March 2, 2014, San Diego, California. Head Neck Pathol 2014;8:1–58.
17 Sobin LH, Gospodarowicz MK, Wittekind CH: TNM Classification of Malignant Tumours, ed 7. New York, Wiley-Liss, 2009.

Patrick J. Bradley, MBA, FRCS
Emeritus Honorary Professor at the School of Medicine, The University of Nottingham
Honorary Consultant Head and Neck Oncological Surgeon, Nottingham University Hospitals
Queens Medical Centre Campus, Derby Road
Nottingham NG7 2UH (UK)

10 Chartwell Grove, Mapperley Plains
Nottingham NG3 5RD (UK)
E-Mail pjbradley@zoo.co.uk

Frequency and Histopathology by Site, Major Pathologies, Symptoms and Signs of Salivary Gland Neoplasms

Patrick J. Bradley

School of Medicine, The University of Nottingham, Nottingham University Hospitals, Queens Medical Centre Campus, Nottingham, UK

Abstract

The frequency distribution of salivary gland neoplasms (SGNs) is, in decreasing order, parotid neoplasms, submandibular gland neoplasms, minor SGNs, and sublingual gland neoplasms. The larger the salivary gland (e.g. parotid), the more likely a neoplasm is benign, and the smaller the gland (e.g. minor salivary gland), the more likely the neoplasm is malignant. The majority of SGNs, benign and/or malignant, irrespective of site, present as a painless swelling or mass. Definitive symptoms and signs of salivary gland malignancy are the presence of named nerve palsy in anatomical proximity to the gland and/or the presence of cervical lymphadenopathy. All discrete major salivary gland masses and non-ulcerated submucosal masses presenting in the head and neck region, irrespective of age, should be investigated, with the aim of excluding an SGN.

© 2016 S. Karger AG, Basel

Introduction

Salivary gland neoplasms (SGNs), from a clinician's perspective, are of two types: benign and malignant. The proportions of benign and malignant neoplasms differ by location (table 1).

In 1953, based on available pathological data, Willis [1], a pathologist from Leeds, England, reported the relative frequencies of SGNs as follows: 'approximately for every 100 parotid tumours, there are likely to be seen about 10 submandibular tumours, 10 tumours of the minor salivary glands (of which about half will be palatal), and only one sublingual tumour'. In 1968, Thackray [2], a pathologist working in London, presented a similar frequency distribution.

Eveson and Cawson [3], both UK pathologists, summarised the experience of the British Salivary Gland Tumour Panel (now disbanded) with 2,356 primary salivary gland tumours (site known) over

Table 1. Tabulation of ratios of benign vs. malignant salivary gland neoplasms by anatomical site in the head and neck in publications that reported complete data

First author [Ref.], year	Cases, n	Anatomical site				Remark
		parotid	submandibular	sublingual	minor salivary	
Thackray [2], 1968		100: 85% B vs. 15% M	10: NR	1: NR	10: 40% B vs. 60% M	Pathologist 'guesstimate'
Eveson [3], 1985	2,356	100: 85% B vs. 15% M (1,498 B: 258 M = 1,356)	11: 63% B vs. 37% M (161 B: 96 M = 57)	<1: 14% B vs. 86% M (1 B: 6 M = 7)	25: 54% B vs. 46% M (180 B: 156 M = 336)	National Pathology Diagnostic Service
Spiro [4], 1986	1,966	100: 75% B vs. 25% M (1,062 B: 354 M = 1,416)	11: 57% B vs. 43% M (92 B: 68 M = 160)	22: 185 B vs. 82% M (72 B: 318 M = 380)		Tertiary Surgical Cancer Service
Li [6], 2008	3,461	100: 68% B vs. 32% M (1,449 B: 675 M = 2,124)	6: 64% B vs. 36% M (240 B: 135 M = 375)	2: 10% B vs. 90% M (5 B: 43 M = 48)	23: 42% B vs. 58% M (393 B: 539 M = 932)	Regional Maxillofacial Surgical Service
Jones [5], 2008	736	100: 73% B vs. 26% M (169 B: 62 M = 231)	30: 61% B vs. 39% M (23 B: 15 M = 38)	1: 25% B vs. 75% M (3 B: 9 M = 12)	200: 62% B vs. 38% M (283 B: 172 M = 455)	Local Regional Pathology Service
Tian [7], 2010	6,982	100: 82% B vs. 16% M (3,500 B: 764 M = 4,264)	13:74% B vs. 26% M (490 B: 173 M = 663)	>1: 5% B vs. 75% M (4 B: 71 M = 75)	46: 38% B vs. 62% M (7,49 B vs. 1,234 M = 1,983)	Surgical specimens from OMFS Regional Centre
Bradley [9], 2013	1,065	100: 91% B vs. 9% M (784 B: 77 M = 861)	11: 76% B vs. 25% M (75 B: 25 M = 100)	0.7: 0% B vs. 100% M (0% B: 6 M = 6)	10: 60% B vs. 40% M (59 B: 39 M = 98)	NHS District Salivary Gland Service
Recommended teaching figure		100: 80% B vs. 20% M	10: 50% B vs. 50% M	1: 5% B vs. 95% M	10: 20% B vs. 80% M	

B = Benign; M = malignant; NR = not recorded; OMFS = oral and maxillofacial surgery; NHS = National Health Service.

Table 2. The more common salivary gland neoplasms, both benign and malignant, presented by anatomical location in the head and neck

	Parotid	Submandibular	Sublingual	Minor Salivary
Benign	PA WT	PA	PA	PA Cystadenoma Canalicular adenoma
Malignant	Mucoepidermoid Adenoid cystic Acinic cell Salivary duct carcinoma Carcinoma ex-pleomorphic adenoma Adenocarcinoma (not otherwise specified)	Adenoid cystic Mucoepidermoid Carcinoma ex-pleomorphic adenoma	Adenoid cystic Mucoepidermoid	Adenoid cystic Mucoepidermoid Polymorphic low-grade adenocarcinoma

a period of approximately 9 years. These cases were referred by pathologists from all over the UK, who had been invited to submit all their salivary gland tumours to this expert panel for confirmation of diagnosis, thus not representing any defined demographic criterion.

In 1986, Spiro [4], at Memorial Sloan Kettering Cancer Center, USA, reported experience with treating patients with SGNs over a 35-year period; of these SGNs, 1,966 were primary lesions, and 841 had been treated previously. Tumours located in the sublingual gland were arbitrarily designated as floor of mouth in origin. This report was composed of surgically treated cases most referred from elsewhere or by the patients themselves, thus not representing any defined demographic area or population.

Jones et al. [5] reported another series from the Department of Oral Pathology, Sheffield, England, between the years 1974 and 2005. They frankly admitted that their series did not accurately support an accurate picture of the true distribution within any defined population. Two large series from China [6, 7] that included several thousands of patients both involved a centralised service for diagnosis and treatment in oral and maxillofacial surgery, with a bias towards minor SGNs and inclusion of only cases post-surgery.

Spiro [8] commented that 'the probability of a malignant diagnosis is <25% in patients with parotid gland tumour, ≈50% with submandibular, >80% in those with minor salivary lesions, and virtually 100% in those few who present with a sublingual lesions. It is important to remember that statistics on the distribution of SGN and the proportion that are malignant usually derive from the tumour registries of large tertiary centres, where there is obvious referral bias. In community hospital setting, virtually all of the salivary gland tumours encountered originate in the parotid and the incidence of malignant tumours is usually lower'.

Bradley and McGurk [9] in the UK reviewed the prevalence of SGNs in a fixed population using the unified computer pathology-based records of two private teaching hospitals for all SGNs, consisting of 1,067 cases (benign and malignant) over a 20-year period (January 1988 to December 2007), with a catchment population of 750,000 persons. The frequency of SGNs confirmed the 100:10:1:10 rule of site distribution, but the benign-to-malignant ratio did not conform to standard teaching. It is recommended that these figures (modified) should be used as a guide when teaching undergraduate medical and nursing students about SGNs.

Salivary Gland Neoplasms in Children

If a solitary mass presents in a major salivary gland of a child, the chances of malignancy are greater than in an adult. Most epithelial tumours occur in children older than 10 years. If malignant, most tumours are low-grade mucoepidermoid carcinoma (MEC) or benign pleomorphic adenoma (PA). Malignant tumours in children younger than 10 tend to be high grade and associated with a poor prognosis. The parotid gland is involved in more than 90% of patients, and until recently, it was reported that >50% of neoplasms were malignant. However, the Nationwide Survey of Danish Pathology Registries reported a series of children (61 cases) over a period of 16 years who had a diagnosis of parotid salivary gland tumours, of which 85% of tumours were benign and 15% were malignant, similar to what is observed in the adult population [10]. Minor SGNs have been reported primarily in the oral cavity, but other head and neck sites have also been recorded [11].

Clinical Summary of the Common Salivary Pathologies (table 1)

Benign Neoplasms
Pleomorphic Adenoma
PA accounts for about 70% of all SGNs in the head and neck. The majority, or 80%, are located in the parotid gland; 10%, in the submandibular gland; and 10%, in the minor salivary gland. Of those that present in the parotid gland, approximately 10% present in the deep lobe (beneath the facial nerve). Palatal tumours are frequently found at the junction of the soft and hard palates and are fixed, rather than mobile, due to the proximity of the underlying mucoperiosteum. Seldom are PAs reported in the sublingual gland. The mean age at presentation is 46 years, but cases have been reported at all ages. PAs usually present as slow-growing, painless masses or swelling.

The small tumours typically form smooth, mobile, firm lumps, but as they become larger, they tend to become more bosselated and may attenuate the overlying skin or mucosa. PAs are circumscribed and encapsulated and are often lobulated, with a fibrous capsule of varying thickness. PAs usually present as solitary lesions, although they may be synchronous or metachronous with other tumours, and usually Warthin tumours (WTs), but also MEC, acinic cell carcinoma (AcCC) and adenoid cystic carcinoma (ACC).

Warthin Tumour
WTs account for 15–20% of SGNs and are the second most common salivary tumour. WTs present only in adults, are more frequent in men, have a strong association with cigarette smoking, and present as a painless, slowly growing mass in the parotid gland. WTs are well-encapsulated lesions usually occurring at the caudal pole of the parotid gland. WTs may presents bilaterally (7–10%), either metachronously (90%) or synchronously (10%) and as multiple lesions within the same gland (13%) [12]. Occasionally, WTs may present with mild pain. Some demonstrate necrosis and present as a diagnostic dilemma in cystic parotid lesions. There appear to be regional, national and racial differences, and correct estimation of the true incidence of WTs is therefore difficult to perform.

Malignant Neoplasms
Mucoepidermoid Carcinoma
MEC is the most common primary SGN in adults and children. The majority present as an asymptomatic swelling or mass. Approximately half of MECs present in the major salivary glands and intraoral sites, and most commonly the palate and buccal mucosa. MECs present in the major salivary glands as firm lesions but are variably movable or fixed. Pain and facial paralysis may be early symptoms. Some MECs have a cystic component and may be misdiagnosed as a mucocoeles when intraorally located. MECs are non-encapsu-

lated neoplasms and are mostly poorly circumscribed. Several grading systems have been described, allowing separation into low-, intermediate- and high-grade disease. The clinical behaviour of a low-grade MEC is 'benign' compared to the intermediate- and high-grade types, which are 'aggressive malignant'. Clinicians have a tendency to consider the intermediate- and high-grade tumours as a group that requires similar treatment. MEC, when present in the submandibular gland, tends to behave 'aggressively', irrespective of its histological grading.

Adenoid Cystic Carcinoma
ACC represents approximately 10% of all epithelial SGNs. ACC makes up 30% of all minor SGNs. The tumour occurs in all age groups, although seldom in children and more frequent in middle-aged and older patients. There is no sex predilection, except for a high incidence in the submandibular gland in women. ACC presents as a slow-growing mass, followed by pain due to perineural and extraparenchymal invasion. Most ACCs are solid and well circumscribed but non-encapsulated. Three define patterns have been described: tubular, cribriform and solid. A recent re-evaluation of a large series suggested that the presence of any solid component is a reliably poor prognostic predictor [13]. Lymph node involvement is uncommon but is most usual at submandibular gland sites [14], and the estimated rate at initial presentation for all stages of ACC is up to 20% [15]. The rate of distant metastasis is estimated to range from 25 to 55% during the course of the disease. The 5-year survival is approximately 35%, but the long-term survival is poorer, with 80–90% of patients dying within 10–15 years. The local recurrence rate has been reported to range from 15 to 85% and is usually a sign of incurability.

Acinic Cell Carcinoma
AcCC affects women more than men at all ages. The parotid gland is involved in >80% of cases (representing 2–4% of all parotid tumours); the intraoral minor salivary glands, in 17%; and the submandibular gland, in 4%. A recent review suggested that most 'non-parotid AcCCs' represent mammary analogue secretory carcinoma, first described in 2010 [16]. This neoplasm usually presents as a slow-growing, solitary, unfixed mass and may present with pain or discomfort. Staging is a better predictor of outcome than histopathological grading is. Most tumours are considered to be of low grade and thus have a favourable outcome. Several factors have been identified to be associated with decreased survival, including male sex, age >45 years, neoplasms larger than 3 cm, and the development of a distant metastasis [17].

Polymorphous Low-Grade Adenocarcinoma
Polymorphous low-grade adenocarcinoma (PLGA) is almost exclusively a minor SGN, but cases arising in the major salivary glands have been reported more recently, although this tumour type has been recognised as a distinct entity since the mid-1980s [18]. This neoplasm presents as a well-circumscribed, but not encapsulated, tumour. Pain and ulceration may be present in 8% of cases. PLGA shares many similar histological patterns with ACC, and cytology may also be confused with that of WTs, PA and sometimes carcinoma ex-pleomorphic adenoma (CXPA). Controversy has arisen as to whether palatally located lesions (low-risk lesions) differ from extrapalatal lesions (high-risk lesions) in terms of the risk of local recurrence and cervical nodal metastasis [18, 19]. It has been estimated that PLGA constitutes 10–20% of malignant intraoral malignancies, second only to MEC.

Carcinoma Ex-Pleomorphic Adenoma
CXPA frequently presents as a painless mass in the parotid gland, and occasionally in the submandibular gland (82%) and in the palatal and intraoral minor salivary glands (18%). Pain may also be a presenting symptom as well as facial

nerve palsy in 30% of patients. The typical picture is a patient presenting with a long-standing parotid mass (many years) that suddenly has increased considerably in size. The tumours are macroscopically poorly circumscribed, and the cut surface often shows haemorrhagic and necrotic areas [20].

Salivary Duct Carcinoma
Salivary duct carcinoma (SDC) is an aggressive SGN that microscopically resembles high-grade ductal carcinoma of the breast, including both in-situ and invasive patterns [21]. It is typically found in older men, most often in the parotid. It can arise de novo or as a malignant component of CXPA. Previously considered rare, SDC is now recognised and accounts for up to 7–10% of malignant SGNs. It presents as a firm, ill-defined mass infiltrating the surrounding gland and soft tissue and is often associated with pain and facial paresis. The majority of cases are Stage III/IV disease. A large proportion of patients present with a T3/T4 tumour (65–80%), and more than 50% of cases present with a palpable cervical metastasis.

Adenocarcinoma (Not Otherwise Specified)
Adenocarcinoma (not otherwise specified) is a salivary gland carcinoma (SGC) that shows glandular or ductal differentiation but lacks the prominence of any of the morphological features that characterise the other, more specific carcinoma types. Thus, the diagnosis is usually one of exclusion. Clinical behaviour is dependent on tumour grading [22].

High-Grade Transformation and 'Dedifferentiation'
High-grade transformation (HGT) is defined as the abrupt transformation of a well-differentiated tumour (described as the conventional tumour) into a tumour with high-grade morphology that lacks the original distinct histological characteristics. HGT has been described in a variety of SGCs, including AcCC, ACC, epithelial-myoepithelial carcinoma, PLGA, low-grade adenocarcinoma, low-grade MEC and hyalinising clear cell carcinoma, although the phenomenon is a rare event [23]. SGC with HGT has been shown to be more aggressive than conventional carcinomas, with a poorer prognosis, accompanied by a higher local recurrence rate and a higher propensity for cervical node metastasis. HGT of SGC can occur either at initial presentation or, less commonly, at the time of recurrence. The most useful tool in identifying the transformed component is a combination of morphological criteria aided by Ki67 expression analysis [24].

Symptoms and Signs

A *sign* is something that can be observed and recognised by a doctor or healthcare professional. A *symptom* is something that only the person experiencing it can feel and know.

Parotid Gland
A slow-growing, painless swelling or mass is the most frequent sign and symptom of both benign and malignant SGNs of the parotid in adults and children. The majority (85%) are located in the superficial/lateral lobe of the parotid; 11% are in the deep lobe, including the parapharyngeal space; and 1% are in the accessory parotid tissue. Benign tumours may grow to enormous sizes, but regardless of the size, the facial nerve is not affected. Pain is seldom a presenting sign, but when present along with a salivary gland malignancy (ACC, SDC, or MEC), it is associated with a poor prognosis. First-bite syndrome has been reported as an initial symptom of ACC and MEC [25, 26]. The duration of the swelling, in years, has been found not to add any reassurance as to whether the mass is benign or malignant. Very large tumour masses are more likely to be malignant than benign. Tumours of the deep lobe are frequently asymptomatic and only detected upon oral exam-

ination or found incidentally following the performance of a CT, MRI, or positron emission tomography scan. Occasionally, a large tumour in the parapharyngeal space may reduce the oropharyngeal airway and present with sleep apnoea or even dyspnoea. Palatal swelling associated with a parotid mass is pathognomonic of a tumour originating in the deep lobe of the parotid. A mass arising in the accessory parotid tissue has an equal chance of being benign or malignant. The diagnosis of a malignant parotid tumour is obvious when a patient presents with a diffuse, large, fixed mass, along with palpable cervical lymph nodes and/or facial nerve palsy. A facial nerve palsy at presentation is noted in 12–15% of cases and is most often associated with ACC, MEC and SDC, and especially high-grade tumours at an advanced stage [26]. Skin ulceration may be present in aggressive high-grade SGN. In this case, one should be wary, as this ulceration may also suggest alternative malignant processes, such as intraparotid lymph node metastasis from a cutaneous skin malignancy or a lymphoma. Another potential confounder is acute inflammation of a WT, presenting with pain and rapid onset of parotid swelling and associated with facial nerve weakness [27].

Submandibular Gland
A painless swelling in the submandibular triangle is the usual symptom and sign of a submandibular SGN. The presence of a painful mass caused by a tumour is very infrequent and is usually associated with sialadenitis. When pain is present, it should be investigated as potentially representing a benign disorder. Malignant lesions usually present as solitary swellings that are firm and lobulated. Attachment to the skin has been reported in less than 25% of cases, but fixation to the deeper structures is more common and can be confirmed by bimanual palpation of the gland intraorally. The only definite sign of malignancy is paralysis of a named nerve (lingual, hypoglossal or marginal mandibular branch of the facial nerve) and/or a fixed submandibular mass [28].

Sublingual Gland
A painless, non-ulcerated submucosal swelling that has been present for less than 12 months is the most usual presentation. Occasionally, the neoplasm will obstruct the submandibular duct and present as submandibular sialadenitis. The majority of lesions at presentation are no more than 3×3 cm. The presence of pain and/or numbness has been reported in up to 50% of patients [29].

Minor Salivary Gland
The majority of SGNs present as asymptomatic, firm submucosal masses in the upper aerodigestive tract. As the majority present in the oral cavity, they can be seen by the patient or observed by clinical examination. The overlying mucosa is seldom ulcerated, unless irritated by the wearing of a denture. Minor SGNs at other sites, such as the nose, paranasal sinus, pharynx and larynx, whether benign and/or malignant, typically present with symptoms of airway restriction or obstruction [see chapter by Bradley and Ferris, this vol., pp. 113–119].

References

1 Willis RA: Epithelial tumours of the salivary glands; in Pathology of Tumours, ed 2. London, Butterworth, 1953, pp 320–348.
2 Thackray AC: Salivary gland tumours – pathology. Proc Roy Soc Med 1968;61: 1089–1092.
3 Eveson JW, Cawson RA: Salivary gland tumours. A review of 2,410 cases with particular reference to histological types, site, age and sex distribution. J Pathol 1985;146:51–58.
4 Spiro RH: Salivary neoplasms: Overview of a 35 year experience with 2,807 patients. Head Neck Surg 1986;8:177–184.

5 Jones AV, Craig GT, Speight PM, et al: The range and demographics of salivary gland tumours diagnosed in a UK population. Oral Oncol 2008;44:407–417.
6 Li LJ, Li Y, Wen YM, et al: Clinical analysis of salivary gland tumour cases in West China in past 50 years. Oral Oncol 2008;44:187–192.
7 Tian Z, Li L, Wang L, et al: Salivary gland neoplasms in oral and maxillofacial regions: a 23-year retrospective study of 6982 cases in an Eastern Chinese population. Oral Maxillofac Surg 2010;39:235–242.
8 Spiro RH: Management of malignant tumours of the salivary glands. Oncology 1998;12:671–683.
9 Bradley PJ, McGurk M: Incidence of salivary gland neoplasms in a define UK population. Br J Oral Maxillofac Surg 2013;51:399–403.
10 Stevens E, Andresen S, Bjørndal K, et al: Tumors in the parotid are not relatively more often malignant in children than in adults. Int J Pediatr Otorhinolaryngol 2015;79:1192–1195.
11 Mehta D, Willging JP: Pediatric salivary gland lesions. Semin Pediatr Surg 2006; 15:76–84.
12 Ethanandan M, Pratt CA, Morrison A, et al: Multiple synchronous and metachronous neoplasms of the parotid gland: the Chicester experience. Br J Oral Maxillofac Surg 2006;44:397–401.
13 van Weert S, van der Waal I, Witte BI, et al: Histopathological grading of adenoid cystic carcinoma of the head and neck: Analysis of currently grading systems and proposal for a simplified grading scheme. Oral Oncol 2015;51:71–76.
14 Bhayani M K, Yener M, EL-Naggar A, et al: Prognosis and risk factors for early-stage adenoid cystic carcinoma of the major salivary glands. Cancer 2012;118: 2872–2878.
15 Lloyd S, Yu JB, Wilson LD, et al: Determinants and patterns of survival in adenoid cystic carcinoma of the head and neck including an analysis of adjuvant radiation therapy. Am J Clin Oncol 2011;34:76–81.
16 Bishop JA, Yonescu R, Batista D, et al: Most non-parotid acinic cell carcinomas represent mammary analogue secretory carcinomas. Am J Surg Pathol 2013;37: 1053–1057,
17 Neskey DM, Klein JD, Hicks S, et al: Prognostic factors associated with decreased survival in patients with acinic cell carcinoma. JAMA Otolaryngol Head Neck Surg 2013;139:1195–1202.
18 Kimple AJ, Austin GK, Shah RN, et al: Polymorphous low-grade adenocarcinoma; A case series and determination of recurrence. Laryngoscope 2014;124: 2714–2719.
19 Seethala RR, Johnson JT, Barnes L, et al: Polymorphous low-grade adenocarcinoma. Arch Otolaryngol Head Neck Surg 2010;136:385–392.
20 Wee DTH, Thomas AA, Bradley PJ: Salivary duct carcinoma: what is already known, and can we improve survival? J Laryngol Otol 2012;126(suppl S2): S2–S6.
21 Huang AT, Tang C, Bell D, et al: Prognostic factors in adenocarcinoma of the salivary glands. Oral Oncol 2015;51: 610–615.
22 Deganello A, Meccariello G, Busoni M, et al: First bite syndrome as presenting symptom of parapharyngeal adenoid cystic carcinoma. J Laryngol Otol 2011; 125:428–431.
23 Nago T: 'Dedifferentiation' and high-grade transformation in salivary gland carcinomas. Head Neck Pathol 2013; 7:S37–S47.
24 Costa FA, Altmani A, Hermsen M: Current concepts on dedifferentiation/high-grade transformation in salivary gland tumors. Pathol Res Int 2011;2011: 325965
25 Diercks GR, Rosow DE, Prasad M, et al: A case of preoperative 'first-bite syndrome' associated with mucoepidermoid carcinoma of the parotid gland. Laryngoscope 2011;121:760–762.
26 Wierzbicka M, Kopec T, Szyfter T, et al: The presence of facial nerve weakness on diagnosis of a parotid gland malignant process. Eur Arch Otorhinolaryngol 2012;269:1177–1182.
27 Mantsopoulos K, Psychogios G, Agaimy A, et al: Inflamed benign tumors of the parotid gland: Diagnostic pitfalls from a potentially misleading entity. Head Neck 2015;37:23–29.
28 Weber RS, Byers RM, Petit B, et al: Submandibular gland tumours, adverse histologic factors and therapeutic implications. Arch Otolaryngol Head Neck Surg 1990;116:1055–1060.
29 Sun G, Yang E, Tang E, et al: The treatment of sublingual gland tumours. Int J Oral Maxillofac Surg 2010;39: 863–868.

Patrick J. Bradley, MBA, FRCS
Emeritus Honorary Professor at the School of Medicine, The University of Nottingham
Honorary Consultant Head and Neck Oncological Surgeon, Nottingham University Hospitals
Queens Medical Centre Campus, Derby Road
Nottingham NG7 2UH (UK)

10 Chartwell Grove, Mapperley Plains
Nottingham NG3 5RD (UK)
E-Mail pjbradley@zoo.co.uk

Molecular Pathology and Biomarkers

Patrick K. Ha[a] · Göran Stenman[b]

[a]Head and Neck Oncologic Surgery, University of California San Francisco, San Francisco, Calif., USA; [b]Sahlgrenska Cancer Center, Department of Pathology, University of Gothenburg and Sahlgrenska University Hospital, Gothenburg, Sweden

Abstract

The field of salivary gland tumor biology is quite broad, given the numerous subtypes of both benign and malignant tumors originating from the major and minor salivary glands. Knowledge about the molecular pathology of these lesions is still limited, and there are few clinically useful diagnostic and prognostic biomarkers. However, recent discoveries of certain key genomic alterations, such as chromosome translocations, copy number alterations, and mutations, provide new insights into the molecular pathogenesis of these lesions and may help to better define them. It is also hoped that this new knowledge can help to guide therapy, but this translation has been somewhat slow to develop, perhaps due to the rarity of these tumors and the lack of large, randomized studies. However, because of the limitations inherent in what surgery and radiation can provide, there is an urgent need for understanding of the mechanisms of carcinogenesis in these tumors individually, so that chemotherapy and/or targeted therapy can be rationally selected.

© 2016 S. Karger AG, Basel

Introduction

For the majority of salivary gland tumors, the etiology remains unknown, with presumably sporadic alterations leading to tumor formation. There are no hereditary syndromes associated with salivary gland neoplasms other than the recently suggested association between germline *BRCA* mutations and salivary gland cancer [1]. Nor are these lesions exposure related, besides the relationship between the benign Warthin tumor and smoking. As such, research into these tumors has not had the benefit of such occurrences that often help to point to critical events in tumorigenesis.

The purpose of this chapter is to focus on the tumor types for which we have significant knowledge of the molecular alterations. Interestingly, there are unique molecularly defining events for many of these tumor types (table 1). Each tumor subtype is reviewed, and what is known about their molecular underpinnings that currently helps to define their characteristics is presented.

Table 1. Recurrent chromosome translocations and gene fusions in salivary gland neoplasms

Diagnosis	Chromosome translocation	Gene fusion	Comments
MEC	t(11;19) t(11;15)	CRTC1-MAML2 CRTC3-MAML2	Found mainly in low- and intermediate-grade MECs
High-grade MEC-like tumors	t(6;15)	EWSR1-POU5F1	
ACC	t(6;9)	MYB-NFIB	Fusion-negative ACCs have MYB gene activation through other mechanisms
HCCC	t(12;22)	EWSR1-ATF1	
CXPA	t(8q12) t(12q14-15)	PLAG1 fusions HMGA2 fusions	Same fusions as in pleomorphic adenomas
MASC	t(11;15)	ETV6-NTRK3	ETV6-NTRK3 is found also in secretory carcinomas of the breast as well as in certain pediatric neoplasms
CAMSG		ARID1A-PRKD1 DDX3X-PRKD1	PRKD1 is also mutated in polymorphous low-grade adenocarcinoma
PA	t(8q12)	CTNNB1-PLAG1 FGFR1-PLAG1 TCEA1-PLAG1 CHCHD7-PLAG1 LIFR-PLAG1	
	t(12q14-15)	HMGA2-NFIB HMGA2-WIF1 HMGA2-FHIT	

Mucoepidermoid Carcinoma

Mucoepidermoid carcinoma (MEC) is the most common of the salivary gland malignancies (≈34% of carcinomas [2]) and is thought to arise from stem/progenitor cells in excretory ducts. Its clinical behavior is known to be dependent on the histology [3]. Tumors with a predominant epithelial/solid component are considered high-grade, whereas those with a predominant cystic/mucoid component are low-grade. The high-grade tumors are known to behave quite aggressively, with a high likelihood of regional and distant metastases, whereas the low-grade tumors are quite indolent in their course.

MECs are characterized by a unique t(11;19) (q21–22;p13) translocation, resulting in a fusion of the two transcriptional coactivators master-mind-like 2 (MAML2) and cyclic AMP (cAMP) response element-binding protein (CREB)-regulated transcription coactivator 1 (CRTC1) [4, 5]. The CRTC1-MAML2 fusion gene encodes a chimeric protein in which the Notch-binding domain of MAML2 is replaced by the CREB-binding domain of CRTC1, or more rarely CRTC3, fused to the transactivation domain of MAML2. The fusion protein activates transcription of cAMP/CREB target genes, including, for example, the epidermal growth factor receptor (EGFR) ligand amphiregulin, leading to activation of EGFR signaling and increased growth and survival of MEC cells [6]. Interestingly, recent studies have suggested that CRTC1-MAML2-positive MEC cells are highly sensitive to inhibition of

EGFR signaling in a xenograft model, suggesting that targeting this pathway may offer a new approach to systemic treatment of patients with advanced, unresectable fusion-positive MECs [6].

Recent studies have convincingly demonstrated that the *CRTC1-MAML2* fusion is a specific and clinically useful biomarker preferentially found in low- and intermediate-grade MECs with a favorable prognosis [7–9]. The fusion may occasionally also be found in high-grade MECs with a poor prognosis, and in these cases, it is often in combination with loss of the *CDKN2A* tumor suppressor gene [10]. However, most high-grade MECs are fusion-negative, indicating that they may represent misclassification of high-grade MEC-like adenocarcinomas (not otherwise specified). A recent genome-wide array comparative genomic hybridization study suggested that MECs may be subclassified into (i) fusion-positive, low- and intermediate-grade MECs with no or few genomic imbalances and a favorable prognosis; (ii) fusion-positive, high-grade MECs, often with *CDKN2A* deletions, multiple genomic imbalances, and a poor prognosis, and (iii) fusion-negative, high-grade MEC-like adenocarcinomas with multiple genomic imbalances and a poor prognosis [8]. Interestingly, previous studies have shown that a subset of the latter tumors has a t(6;22)(p21;q12) translocation, resulting in an *EWSR1-POU5F1* gene fusion [11]. Taken together, these studies show that the spectrum of MEC-like tumors is broad and heterogeneous both histologically and genetically.

Adenoid Cystic Carcinoma

Adenoid cystic carcinoma (ACC) arises from stem/progenitor cells in intercalated ducts and may have quite a varied clinical course. It is the second most common salivary gland carcinoma, constituting approximately 22% of malignant salivary tumors [2]. While these tumors have a generally slow growth rate, they do have a poor long-term prognosis, with a high propensity for perineural invasion and distant metastases. There is currently no effective treatment available for patients with advanced recurrent and/or metastatic disease.

ACC is characterized by a t(6;9)(q22–23;p23–24) chromosomal translocation, resulting in a fusion involving the *MYB* oncogene and the transcription factor gene *NFIB* [12]. The *MYB-NFIB*-encoded fusion oncoprotein activates transcription of MYB targets of key importance for oncogenic transformation [9]. *MYB* activation due to gene fusion or other mechanisms is found in more than 80% of head and neck ACCs and is a novel diagnostic biomarker for this disease [9, 13, 14]. There is currently no evidence to suggest that the fusion has prognostic significance. However, since *MYB* is an oncogenic driver in ACC, it is also a potential therapeutic target. At present, it is not known whether MYB-negative ACCs are true ACCs or whether they represent a subset of tumors with an ACC-like morphology, such as, for example, polymorphous low-grade adenocarcinoma (PLGA) or basal cell adenocarcinoma.

Genomic profiling studies have shown that ACC generally has a rather quiet genome, with few copy number alterations compared to other high-grade carcinomas, consistent with the notion that the *MYB-NFIB* fusion is the major oncogenic event in ACC [12, 15]. Deletions involving 12q, 6q, 9p, 11q, 14q, 1p, and 5q and gains involving 1q and 22q are most common. Losses of 1p, 6q, and 15q are specifically associated with high-grade tumors, whereas loss of 14q is exclusively seen in grade I tumors. There is now clear evidence to suggest that deletion of 1p36.33-p35.3 is a biomarker associated with high-grade tumors with a poor prognosis [12, 16].

Using next-generation sequencing, the complete exomes of more than 80 ACCs have been sequenced. Stephens et al. [17] reported a relatively low average mutational rate of 13 per sample, with a range of 2–35 mutations per sample.

The most common mutations were in *NOTCH1/2* (3/24 cases), *SPEN* (5/24 cases), and *FGF2* (3/24 cases). There was additionally a cluster of mutations found in genes involved in histone modification and chromatin remodeling. Similar findings of a low mutational rate (22 somatic mutations per sample on average, with a range of 1–36) were also identified by Ho et al. [15], along with similar identification of mutations in chromatin-regulating genes in 35% of cases. Additionally, there were mutations within the *FGF/PI3* kinase pathway in 30% of samples as well as in the *NOTCH1* pathway in 13% of the cases.

A third sequencing study was performed in a cohort of patients with recurrent and metastatic ACCs, looking specifically at cancer-related genes [18]. Alterations that could be targeted by specific therapies were evaluated; 44 actionable alterations were found in the cohort of 28 tumors, and 12/28 (43%) of patients had at least one mutation. Therefore, it was postulated that the use of DNA sequencing can help to guide therapy in those patients who otherwise may have no other known treatment option. Taken together, the results demonstrate that the mutational rate of any one particular gene in ACC is quite low, suggesting that, apart from the *MYB-NFIB* gene fusion, there are numerous alterations that likely must occur for the cancer to arise.

KIT is a tyrosine kinase thought to have oncogenic effects with many possible downstream targets that promote growth. There appears to be a predominance of expression in the solid variant of ACC [19], and an overwhelming number of studies has identified overexpression in 78–100% of cases [20–27]. Mutations in *KIT* in ACC are very rare [15, 17].

EGFR is a known oncogene that encodes a cell surface protein with positive growth effects upon heterodimerization. In other tumor types, such as head and neck squamous cell carcinoma, there are known activating mutations in exons 19 and 21, with oncogenic effects. In ACC, there appears to be overexpression in up to 75–85% of tumors [28], but the presence of these mutations is rare [29]. Furthermore, overexpression is believed to not be associated with gene amplification [30].

Salivary Duct Carcinoma

Salivary duct carcinoma (SDC) is an aggressive cancer with a high risk of regional and distant metastases and a 5-year overall survival of 43% [31]. Interestingly, it has a similar or identical histological appearance to that found in invasive ductal breast cancer. Thus, this carcinoma shows frequent *HER2 (ERBB2)* overexpression, occurring in >80% of tumors [32, 33] along with gene amplification [34], though it is not believed to be a predictor of survival [23]. The androgen receptor has also been identified to be highly expressed in SDC [35], but not due to gene amplification [14]. EGFR has also been found to be overexpressed in at least half of SDCs [36], although it remains to be seen whether this is a viable therapeutic target. Mutations in *PIK3CA* have been identified in 20–24% of SDCs [37], suggesting that this pathway could be a potential target for therapy.

Mammary Analog Secretory Carcinoma

Mammary analog secretory carcinoma (MASC) is a relatively newly described salivary gland malignancy that predominantly affects the parotid gland [38]. This carcinoma was initially confused with acinic cell carcinoma, low-grade cystadenocarcinoma, or MEC, but the presence of a t(12;15)(p13;q25) translocation and *ETV6-NTRK3* gene fusion, analogous to secretory carcinoma of the breast, identified MASC as a distinct entity [38–40]. The *ETV6-NTRK3* fusion is a hallmark of MASC [41]. The clinical behavior of MASC appears to be similar to that of acinic cell carcinoma, wherein it can behave aggressively at times but typically has a relatively indolent course [40, 41].

Pleomorphic Adenoma

Pleomorphic adenoma (PA) is the most common subtype of salivary gland neoplasm. It is a benign tumor that, if treated adequately, has an excellent prognosis. Histologically, PAs show a pronounced morphological diversity, with epithelial and myoepithelial cells forming a variety of patterns in an often fibrous, myxoid or condroid stroma. PA was the first benign human tumor that was shown to be characterized by tumor type-specific chromosomal translocations [42]. Subsequent studies revealed that PAs are frequently characterized by recurrent translocations preferentially affecting 8q12 (>50%) and 12q14-q15 (≈15%) [43]. These rearrangements result in gene fusions affecting the transcription factor genes *PLAG1* and *HMGA2* (table 1) [43]. These fusions have not been encountered in any other type of salivary gland tumor and may therefore be useful as biomarkers to differentiate PA from its mimics, including, for example, ACC and PLGA. Previous studies have also indicated that PAs with *PLAG1* gene fusions often have activation of HRAS [44]. Gene amplifications and overexpression of genes in 12q13-q15, including *HMGA2*, *MDM2*, *CDK4*, *TSPAN31*, and *GLI1*, are found in a subset of PAs [45]. *TP53* mutations are infrequent in PA but are common in carcinoma ex-PA (CXPA) [45].

Carcinoma Ex-Pleomorphic Adenoma

CXPA is a carcinoma arising within a preexisting PA. CXPA constitutes approximately 12% of all salivary gland carcinomas. The malignant component is often an adenocarcinoma (not otherwise specified) or an undifferentiated carcinoma but may be any other histological subtype. CXPAs that are noninvasive or minimally invasive have an excellent prognosis, whereas invasive tumors are usually high-grade, aggressive lesions with a poor prognosis.

CXPAs are characterized by the genetic alterations that are typical of PAs: that is, translocations and gene fusions involving *PLAG1* and *HMGA2* (table 1). In addition, subsets of these tumors have amplifications of *HMGA2* and *MDM2* in 12q14–q15, mutations of *TP53*, and/or amplification of *HER2* as molecular biomarkers of malignant transformation [9, 45]. CXPAs with *HER2* amplification are mainly SDCs-ex-PAs. Interestingly, there are anecdotal reports showing that these patients may benefit from anti-HER2 treatment with trastuzumab [9]. Other alterations associated with adenoma-carcinoma progression are deletions involving 5q23.2-q31.2 and gain/amplification of the *PLAG1* and *MYC* oncogenes [45].

Hyalinizing Clear Cell Carcinoma

Hyalinizing clear cell carcinoma (HCCC) is a rare, low-grade minor salivary gland carcinoma characterized by clear cells growing in nests and cords in a hyalinized stroma [46]. The prognosis is excellent, with only occasional metastatic spread. HCCC shows histological overlap with other salivary neoplasms, such as myoepithelial carcinoma, epithelial-myoepithelial carcinoma, and MEC. Recent studies have demonstrated that HCCC is frequently characterized by a t(12;22)(q13;q12) translocation, resulting in an *EWSR1-ATF1* gene fusion [46]. *EWSR1* gene rearrangements were recently also detected in clear cell variants of aggressive myoepithelial carcinomas as well as in clear cell odontogenic carcinomas [47, 48]. These findings, together with the original identification of *EWSR1-ATF1* fusions in soft tissue clear cell sarcomas [49], provide evidence for a unifying concept for several tumor types with clear cell morphology.

Polymorphous Low-Grade Adenocarcinoma and Cribriform Adenocarcinoma of Minor Salivary Glands

PLGA and cribriform adenocarcinoma of minor salivary glands (CAMSG) are low-grade carcinomas that mainly occur in the oral cavity and oropharynx, respectively. There is some controversy as to whether they are histogenetically separate entities or variants along one spectrum. PLGAs were recently shown to have activating *PRKD1* hotspot mutations (p.Glu710Asp) in 73% of cases [50]. In contrast, *PRKD1* mutations were not detected in any other subtypes of salivary neoplasms analyzed (n = 299), indicating that *PRKD1* mutations define a subset of PLGA and may be used as a diagnostic biomarker to differentiate PLGA from other mimics. Moreover, *PRKD1* mutations were significantly associated with the metastasis-free survival of patients with malignant salivary gland tumors. Interestingly, classical CAMSG was recently shown to have rearrangements of *PRKD1*, *PRKD2* or *PRKD3* in 75% of the cases, including *ARID1A-PRKD1* and *DDX3X-PRKD1* gene fusions [51]. Taken together, these findings support the concept of a possible shared pathogenesis for PLGA and CAMSG.

Conclusion

It is evident that our knowledge of the molecular pathogenesis of salivary gland tumors is steadily growing, with the identification of novel unique molecular alterations in specific tumor types. It is hoped that further discoveries will lead to improved diagnosis, better prognostication, and additional treatment options for patients with these tumors.

References

1. Shen TK, Teknos TN, Toland AE, et al: Salivary gland cancer in BRCA-positive families: a retrospective review. JAMA Otolaryngol Head Neck Surg 2014;140:1213–1217.
2. Spiro RH: Salivary neoplasms: overview of a 35-year experience with 2,807 patients. Head Neck Surg 1986;8:177–184.
3. Chen MM, Roman SA, Sosa JA, et al: Histologic grade as prognostic indicator for mucoepidermoid carcinoma: a population-level analysis of 2,400 patients. Head Neck 2014;36:158–163.
4. Enlund F, Behboudi A, Andren Y, et al: Altered Notch signaling resulting from expression of a WAMTP1-MAML2 gene fusion in mucoepidermoid carcinomas and benign Warthin's tumors. Exp Cell Res 2004;292:21–28.
5. Tonon G, Modi S, Wu L, et al: t(11;19)(q21;p13) translocation in mucoepidermoid carcinoma creates a novel fusion product that disrupts a Notch signaling pathway. Nat Genet 2003;33:208–213.
6. Chen Z, Chen J, Gu Y, et al: Aberrantly activated AREG-EGFR signaling is required for the growth and survival of CRTC1-MAML2 fusion-positive mucoepidermoid carcinoma cells. Oncogene 2014;33:3869–3877.
7. Behboudi A, Enlund F, Winnes M, et al: Molecular classification of mucoepidermoid carcinomas-prognostic significance of the MECT1-MAML2 fusion oncogene. Genes Chromosomes Cancer 2006;45:470–481.
8. Jee KJ, Persson M, Heikinheimo K, et al: Genomic profiles and CRTC1-MAML2 fusion distinguish different subtypes of mucoepidermoid carcinoma. Mod Pathol 2013;26:213–222.
9. Stenman G, Persson F, Andersson MK: Diagnostic and therapeutic implications of new molecular biomarkers in salivary gland cancers. Oral Oncol 2014;50:683–690.
10. Tirado Y, Williams MD, Hanna EY, et al: CRTC1/MAML2 fusion transcript in high grade mucoepidermoid carcinomas of salivary and thyroid glands and Warthin's tumors: implications for histogenesis and biologic behavior. Genes Chromosomes Cancer 2007;46:708–715.
11. Moller E, Stenman G, Mandahl N, et al: POU5F1, encoding a key regulator of stem cell pluripotency, is fused to EWSR1 in hidradenoma of the skin and mucoepidermoid carcinoma of the salivary glands. J Pathol 2008;215:78–86.
12. Persson M, Andren Y, Moskaluk CA, et al: Clinically significant copy number alterations and complex rearrangements of MYB and NFIB in head and neck adenoid cystic carcinoma. Genes Chromosomes Cancer 2012;51:805–817.
13. Brill LB 2nd, Kanner WA, Fehr A, et al: Analysis of MYB expression and MYB-NFIB gene fusions in adenoid cystic carcinoma and other salivary neoplasms. Mod Pathol 2011;24:1169–1176.

14 Mitani Y, Rao PH, Maity SN, et al: Alterations associated with androgen receptor gene activation in salivary duct carcinoma of both sexes: potential therapeutic ramifications. Clin Cancer Res 2014;20:6570–6581.

15 Ho AS, Kannan K, Roy DM, et al: The mutational landscape of adenoid cystic carcinoma. Nat Genet 2013;45:791–798.

16 Rao PH, Roberts D, Zhao YJ, et al: Deletion of 1p32-p36 is the most frequent genetic change and poor prognostic marker in adenoid cystic carcinoma of the salivary glands. Clin Cancer Res 2008;14:5181–5187.

17 Stephens PJ, Davies HR, Mitani Y, et al: Whole exome sequencing of adenoid cystic carcinoma. J Clin Invest 2013;123: 2965–2968.

18 Ross JS, Wang K, Rand JV, et al: Comprehensive genomic profiling of relapsed and metastatic adenoid cystic carcinomas by next-generation sequencing reveals potential new routes to targeted therapies. Am J Surg Pathol 2014; 38:235–238.

19 Bell D, Roberts D, Kies M, et al: Cell type-dependent biomarker expression in adenoid cystic carcinoma: biologic and therapeutic implications. Cancer 2010;116:5749–5756.

20 Aslan DL, Oprea GM, Jagush SM, et al: c-kit expression in adenoid cystic carcinoma does not have an impact on local or distant tumor recurrence. Head Neck 2005;27:1028–1034.

21 Locati LD, Perrone F, Losa M, et al: Treatment relevant target immunophenotyping of 139 salivary gland carcinomas (SGCs). Oral Oncol 2009;45:986–990.

22 Edwards PC, Bhuiya T, Kelsch RD: C-kit expression in the salivary gland neoplasms adenoid cystic carcinoma, polymorphous low-grade adenocarcinoma, and monomorphic adenoma. Oral Surg Oral Med Oral Pathol Oral Radiol Endod 2003;95:586–593.

23 Ettl T, Schwarz S, Kleinsasser N, et al: Overexpression of EGFR and absence of C-KIT expression correlate with poor prognosis in salivary gland carcinomas. Histopathology 2008;53:567–577.

24 Freier K, Flechtenmacher C, Walch A, et al: Differential KIT expression in histological subtypes of adenoid cystic carcinoma (ACC) of the salivary gland. Oral Oncol 2005;41:934–939.

25 Holst VA, Marshall CE, Moskaluk CA, et al: KIT protein expression and analysis of c-kit gene mutation in adenoid cystic carcinoma. Mod Pathol 1999;12:956–960.

26 Jeng YM, Lin CY, Hsu HC: Expression of the c-kit protein is associated with certain subtypes of salivary gland carcinoma. Cancer Lett 2000;154:107–111.

27 Mino M, Pilch BZ, Faquin WC: Expression of KIT (CD117) in neoplasms of the head and neck: an ancillary marker for adenoid cystic carcinoma. Mod Pathol 2003;16:1224–1231.

28 Vered M, Braunstein E, Buchner A: Immunohistochemical study of epidermal growth factor receptor in adenoid cystic carcinoma of salivary gland origin. Head Neck 2002;24:632–636.

29 Dahse R, Driemel O, Schwarz S, et al: Epidermal growth factor receptor kinase domain mutations are rare in salivary gland carcinomas. Br J Cancer 2009;100: 623–625.

30 Vidal L, Tsao MS, Pond GR, et al: Fluorescence in situ hybridization gene amplification analysis of EGFR and HER2 in patients with malignant salivary gland tumors treated with lapatinib. Head Neck 2009;31:1006–1012.

31 Johnston ML, Huang SH, Waldron JN, et al: Salivary duct carcinoma: treatment, outcomes, and patterns of failure. Head Neck 2015, Epub ahead of print.

32 Glisson B, Colevas AD, Haddad R, et al: HER2 expression in salivary gland carcinomas: dependence on histological subtype. Clin Cancer Res 2004;10:944–946.

33 Skalova A, Starek I, Kucerova V, et al: Salivary duct carcinoma – a highly aggressive salivary gland tumor with HER-2/neu oncoprotein overexpression. Pathol Res Pract 2001;197:621–626.

34 Dagrada GP, Negri T, Tamborini E, et al: Expression of HER-2/neu gene and protein in salivary duct carcinomas of parotid gland as revealed by fluorescence in-situ hybridization and immunohistochemistry. Histopathology 2004;44:301–302.

35 Fan CY, Melhem MF, Hosal AS, et al: Expression of androgen receptor, epidermal growth factor receptor, and transforming growth factor alpha in salivary duct carcinoma. Arch Otolaryngol Head Neck Surg 2001;127:1075–1079.

36 Williams MD, Roberts D, Blumenschein GR Jr, et al: Differential expression of hormonal and growth factor receptors in salivary duct carcinomas: biologic significance and potential role in therapeutic stratification of patients. Am J Surg Pathol 2007;31:1645–1652.

37 Ku BM, Jung HA, Sun JM, et al: High-throughput profiling identifies clinically actionable mutations in salivary duct carcinoma. J Transl Med 2014;12:299.

38 Skalova A, Vanecek T, Sima R, et al: Mammary analogue secretory carcinoma of salivary glands, containing the ETV6-NTRK3 fusion gene: a hitherto undescribed salivary gland tumor entity. Am J Surg Pathol 2010;34: 599–608.

39 Bishop JA, Yonescu R, Batista DA, et al: Cytopathologic features of mammary analogue secretory carcinoma. Cancer Cytopathol 2013;121:228–233.

40 Chiosea SI, Griffith C, Assaad A, et al: Clinicopathological characterization of mammary analogue secretory carcinoma of salivary glands. Histopathology 2012;61:387–394.

41 Bishop JA, Yonescu R, Batista D, et al: Utility of mammaglobin immunohistochemistry as a proxy marker for the ETV6-NTRK3 translocation in the diagnosis of salivary mammary analogue secretory carcinoma. Hum Pathol 2013; 44:1982–1988.

42 Mark J, Dahlenfors R, Ekedahl C, et al: The mixed salivary gland tumor – a normally benign human neoplasm frequently showing specific chromosome abnormalities. Cancer Genet Cytogenet 1980;2:231–241.

43 Stenman G: Fusion oncogenes in salivary gland tumors: molecular and clinical consequences. Head Neck Pathol 2013;7(suppl 1):S12–S19.

44 Stenman G, Sandros J, Mark J, et al: High p21RAS expression levels correlate with chromosome 8 rearrangements in benign human mixed salivary gland tumors. Genes Chromosomes Cancer 1989;1:59–66.

45 Persson F, Andren Y, Winnes M, et al: High-resolution genomic profiling of adenomas and carcinomas of the salivary glands reveals amplification, rearrangement, and fusion of HMGA2. Genes Chromosomes Cancer 2009;48: 69–82.

46 Antonescu CR, Katabi N, Zhang L, et al: EWSR1-ATF1 fusion is a novel and consistent finding in hyalinizing clear-cell carcinoma of salivary gland. Genes Chromosomes Cancer 2011;50:559–570.

47 Bilodeau EA, Weinreb I, Antonescu CR, et al: Clear cell odontogenic carcinomas show EWSR1 rearrangements: a novel finding and a biological link to salivary clear cell carcinomas. Am J Surg Pathol 2013;37:1001–1005.

48 Skalova A, Weinreb I, Hyrcza M, et al: Clear cell myoepithelial carcinoma of salivary glands showing EWSR1 rearrangement: molecular analysis of 94 salivary gland carcinomas with prominent clear cell component. Am J Surg Pathol 2015;39:338–348.

49 Zucman J, Delattre O, Desmaze C, et al: EWS and ATF-1 gene fusion induced by t(12;22) translocation in malignant melanoma of soft parts. Nat Genet 1993;4: 341–345.

50 Weinreb I, Piscuoglio S, Martelotto LG, et al: Hotspot activating PRKD1 somatic mutations in polymorphous low-grade adenocarcinomas of the salivary glands. Nat Genet 2014;46:1166–1169.

51 Weinreb I, Zhang L, Tirunagari LM, et al: Novel PRKD gene rearrangements and variant fusions in cribriform adenocarcinoma of salivary gland origin. Genes Chromosomes Cancer 2014;53: 845–856.

Patrick Ha, MD, FACS
Professor of Otolaryngology, Chief of Head and Neck Oncologic Surgery
University of California San Francisco
550 16th Street, Box 3213, Mission Hall, 4th Floor
San Francisco, CA 94158 (USA)
E-Mail Patrick.Ha@ucsf.edu

Evaluation: Imaging Studies

Marinos Kontzialis[a] · Christine M. Glastonbury[b] · Nafi Aygun[a]

[a]Division of Neuroradiology, Russell H. Morgan Department of Radiology and Radiological Science, The Johns Hopkins University School of Medicine, Baltimore, Md., and [b]Department of Radiology and Biomedical Imaging, University of California San Francisco, San Francisco, Calif., USA

Abstract

The malignant or benign nature of a salivary gland (SG) tumor can be predicted with reasonably high accuracy by imaging. There is some overlap between the imaging findings of benign and malignant tumors, particularly for low-grade malignancies, and tissue diagnosis remains necessary for definitive diagnosis. Magnetic resonance imaging is the modality of choice for the evaluation of salivary neoplasms, as it allows for delineation of local infiltration, perineural spread and intracranial extension. This review will focus on the advanced imaging techniques that help to characterize SG tumors. A brief overview of the conventional imaging features of SG neoplasms is necessary before a discussion of the advanced imaging methods.

© 2016 S. Karger AG, Basel

Conventional Imaging

Ultrasound

Ultrasound is safe, accessible and inexpensive. It can evaluate the submandibular gland and superficial parotid masses and guide fine needle aspiration (FNA). However, ultrasound is operator dependent, and it cannot effectively evaluate the deep lobe of the parotid gland.

Computerized Tomography

Computerized tomography (CT) provides a reasonably high rate of sensitivity for the detection of salivary gland (SG) neoplasms. However, it cannot define the tumor extent and differentiate benign from malignant masses consistently and with high accuracy [1].

Magnetic Resonance Imaging

Magnetic resonance imaging (MRI) has the greatest ability to detect and characterize SG tumors. Benign tumors typically have smooth and well-defined borders, hyperintense signal on T2-weighted (T2W) images and are often superficially located when in the parotid gland (fig. 1). Malignant tumors are more variable in their appearance, depending on their pathologic grade. High-grade malignant tumors typically exhibit ill-defined or infiltrative contours and have lower T2 signal intensity. Low-grade malignant

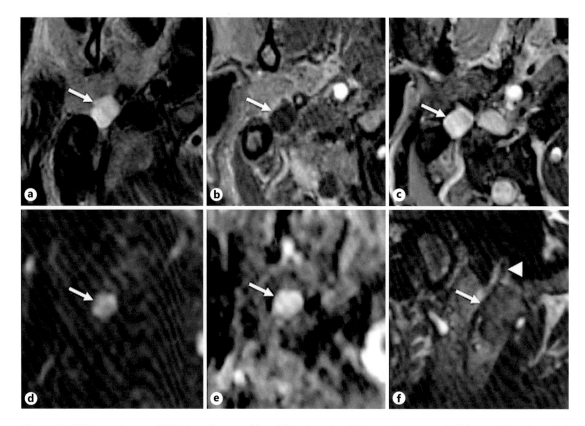

Fig. 1. PA. A T2 hyperintense (**a**), T1 hypointense (**b**), avidly enhancing (**c**), hyperintense on DWI b1000 (**d**) and on ADC map (**e**), well-circumscribed lesion (arrows) with a high ADC value (mean ADC = 2.159×10^{-3} mm^2/s) is centered just inferiorly to the right stylomastoid foramen. On precontrast sagittal constructive interference in steady state (**f**), notice cranial nerve VII exiting the stylomastoid foramen (arrowhead) and extending inferiorly adjacent to the PA (arrow).

tumors can mimic benign neoplasms, with smooth, sharply circumscribed borders, although solid low-grade tumors do not typically have the high T2 signal of pleomorphic adenoma (PA). In general, however, these features are not specific enough to allow a definitive diagnosis [2]. Cystic changes may be present in benign as well as malignant neoplasms, and PAs frequently show heterogeneous enhancement, which may be mistaken for necrosis, particularly when large. Thus, MRI is imperfect for complete characterization, and tissue diagnosis is usually required (fig. 2) [3, 4]. MRI does, however, allow for evaluation of the complete extent of a mass, perineural tumor spread and nodal metastases. The relationship of a parotid gland tumor with the facial nerve is not always easily elucidated.

A capsule is seen around most benign and some malignant parotid gland tumors. An irregular, thick and avidly enhancing capsule-like enhancement is suggestive of malignancy (fig. 2) [5, 6]. Involvement of the deep parotid lobe or both lobes should raise suspicion for malignancy [3]. The single best MRI finding suggestive of malignancy is ill-defined margins on contrast-enhanced images, achieving 70% sensitivity and 73% specificity when combined with T2 hypoin-

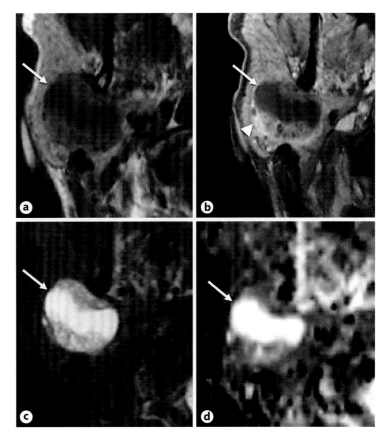

Fig. 2. Acinic cell carcinoma. A right parotid mass (arrows) with well-circumscribed margins, hypointense signal on T1 (**a**) and hyperintense signal on T2W images (**c**). There is central cystic component. The solid component of the mass enhances (**b**), and there is thick capsule-like enhancement posterolaterally (arrowhead in **b**). The solid component of the mass demonstrates a low ADC value (mean ADC = 1.090×10^{-3} mm^2/s), which is suggestive of malignancy.

tensity (fig. 3) [3]. Additional findings predictive of parotid malignancy include diffuse growth; infiltration of the subcutaneous tissues, skin, the masticator space and the parapharyngeal space; lymphadenopathy; and perineural spread (PNS) [3].

Tumor Staging

Staging of SG malignancies is based on tumor size, extraglandular spread, and local extent [1]. A T1 tumor measures less than 2 cm, and a T2 tumor measures between 2 and 4 cm. T3 lesions measure more than 4 cm or demonstrate extraglandular spread. T4a lesions include tumors with skin, mandible, external auditory canal, and/or facial nerve invasion. T4b lesions invade the skull base and/or the pterygoid plates and/or encase the internal carotid artery. Of note, the staging system does not factor in the histologic type and grade of tumors, which are often more important than the initial stage of a tumor in prognostication.

Perineural Spread

PNS refers to retrograde or antegrade spread of tumor from the primary site along a nerve, which is evident on imaging (fig. 4), in contrast to the histologically identified perineural invasion,

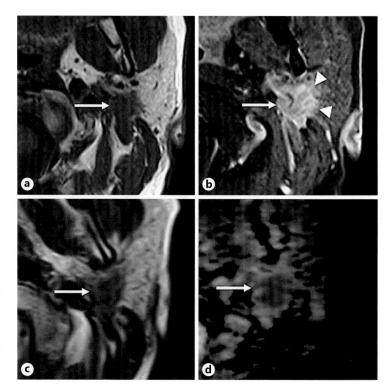

Fig. 3. High-grade salivary duct carcinoma. T1 hypointense (**a**), T2 hypointense (**c**), enhancing lesion (arrows) with spiculated and ill-defined margins (arrowheads in **b**). The lesion has a very low ADC value (mean ADC = 0.821×10^{-3} mm^2/s).

which occurs at the primary site [7]. Although not included in staging schemes, both PNS and perineural invasion predict worse prognosis. The facial nerve and the auriculotemporal branch of the mandibular division of the trigeminal nerve are the nerves most commonly affected by parotid malignancies. In minor SG (MSG) malignant tumors of the palate and in sublingual and submandibular gland tumors, the palatine nerves and the inferior alveolar nerve, respectively, are primarily involved. PNS can cause skip lesions, so tracing the entire course of the nerves is necessary when searching for PNS. On CT, there might be foraminal enlargement and bone destruction [7]. On MRI, signs of PNS include effacement of normal fat planes around nerves on T1-weighted (T1W) images and abnormal nerve thickening and enhancement following contrast administration. Because PNS is often very limited, it is best visualized by high-resolution MRI (HR-MRI), and positron emission tomography (PET) often fails to identify it.

Posttreatment Surveillance

Contrast-enhanced MRI is the method of choice in the posttreatment setting. CT's sensitivity and specificity are low. PET-CT is unrivaled for distant metastases but is not superior to MRI for locoregional recurrences [8, 9]. Diffuse, poorly marginated enhancement and T2 hyperintense signal are common in the parotid bed in the postsurgical setting and can be mistaken for residual or recurrent tumor [1, 10]. The recurrence of PA has a very distinctive appearance that includes clusters of multiple rounded masses scattered throughout the operative bed [11,

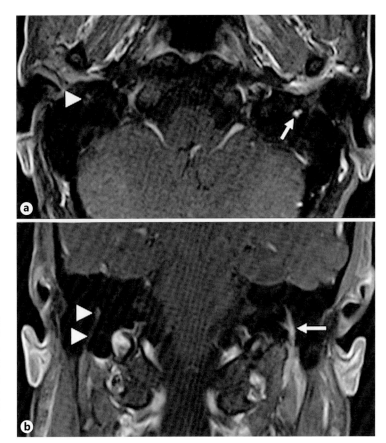

Fig. 4. PNS. Axial (**a**) and coronal (**b**) fat-suppressed postcontrast T1W images demonstrate asymmetric enhancement in the mastoid segment of the left facial nerve (arrows) in a patient with high-grade parotid malignancy. Compare with the normal contralateral facial nerve (arrowheads).

12]. Recurrent PA has a very bright signal, and fat-suppressed T2W and short tau inversion recovery sequences are particularly helpful in defining the extent of the recurrent tumor. Comparison with the presurgical scan and the signal characteristics of the primary neoplasm, the time interval from surgery, review of the operative and pathology reports, and close imaging follow-up are helpful in differentiating postoperative fibrosis and granulation tissue from tumor recurrence. A nodular lesion that increases in size and follows the signal of the resected neoplasm is usually compelling evidence of residual or recurrent disease. Having a low threshold for FNA biopsy of suspicious lesions is imperative for malignant tumors. Careful scrutiny of the facial and trigeminal nerves is important, as PNS may be the sole manifestation of tumor recurrence.

Advanced Imaging Techniques

High-Resolution Magnetic Resonance Imaging
HR-MRI with a slice thickness of 0.5 mm and isotropic voxels provides remarkably detailed images of the SG. The local extent of SG tumors outside the gland is evaluated with great precision, particularly at the skull base, where conventional imaging techniques have shortcomings. Three-dimensional constructive interference in steady state (3D-CISS) images allow visualization of the

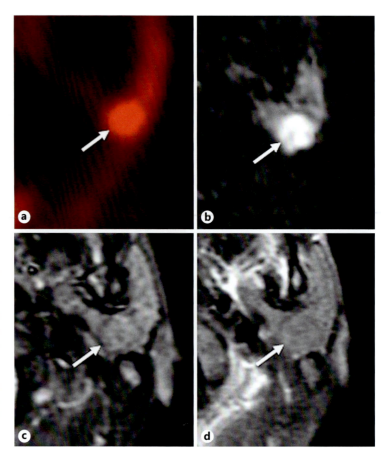

Fig. 5. WT presenting as incidental focal parotid uptake on PET. A hypermetabolic lesion (**a**; arrows) demonstrates hyperintense DWI signal (**b**) and restricted diffusion (mean ADC = 0.800×10^{-3} mm^2/s) and is barely discernible from and is isointense to the normal parotid gland, with a faint hypointense rim on T2 (**c**) and precontrast T1 (**d**). The mass remains isointense to the parotid gland on postcontrast imaging (not shown).

intraparotid facial nerve and its branches up to the third order (fig. 1f) [13]. With HR-MRI, PNS of malignancy can be identified with higher sensitivity, and its extent can be defined with greater accuracy. The relationship of parotid tumors with the facial nerve can be determined preoperatively, although the effectiveness of such practice in decreasing postoperative facial weakness has not been established.

Diffusion-Weighted Imaging
Diffusion-weighted imaging (DWI) MRI depicts the Brownian motion of water molecules in biological tissues [14]. A hypercellular tumor with densely packed cells and more cell membranes is expected to present a greater impediment to diffusion and thus a lower apparent diffusion coefficient (ADC) [14, 15]. DWI has been applied to the head and neck to detect and characterize tumors, to detect lymph nodes, to monitor treatment response, and to differentiate recurrent tumor from posttreatment changes [14]. PAs (fig. 1) have higher ADC values than malignancies (fig. 2, 3) and Warthin tumors (WTs; fig. 5), which contain lymphoid tissue [16, 17]. In the largest series to date, which included 136 patients, pleomorphic and myoepithelial adenomas had the highest ADC values, allowing distinction from SG malignancies. However, definite differentiation of benign from malignant lesions was

not possible due to the overlap in the ADC values of WTs and malignancies (fig. 3, 5) [18]. Despite the large number of patients studied, no lymphomas, which are known to have markedly decreased ADC values that overlap with the values of Warthin and malignant tumors, were included [19, 20]. When applying DWI clinically, it is important to keep in mind that ADC values will vary based on the b values (a maximum b value of 800 or 1,000 is typically used), the DWI sequence and the MRI system used and that reported ADC values are not directly transferable between institutions [15, 21]. DWI has shown promising results in differentiating posttreatment changes from recurrent tumor in head and neck squamous cell carcinoma [22]. DWI's role in surveillance of SG tumors after treatment is an area of active investigation.

Dynamic Contrast-Enhanced Magnetic Resonance Imaging

Dynamic contrast-enhanced (DCE)-MRI has shown promise in the characterization of SG neoplasms [23]. Following contrast administration, multiple T1-weighted images of the salivary tumor are obtained for several minutes to monitor the uptake and washout of contrast. The time of peak enhancement correlates with the microvessel count and tends to be short when the microvessel count is high. Washout depends on the cellularity-stromal grade, with cellular neoplasms having faster washout [23]. In brief, PAs tend to demonstrate progressive enhancement (low microvessel count and cellularity-stromal grade) [23, 24]. WTs demonstrate rapid enhancement and washout (high microvessel count and cellularity-stromal grade) [23, 24]. Malignant neoplasms demonstrate rapid enhancement, but washout tends to be relatively slower than in WTs (high microvessel count and lower cellularity-stromal grade) [23, 24]. As with DWI, DCE-MRI alone cannot differentiate benign from malignant tumors due to overlapping patterns of enhancement; for instance, lymphomas demonstrate early enhancement and high washout, similar to WTs [24, 25]. The addition of ADC values appears to increase accuracy when combined with MRI perfusion, especially for tumors demonstrating washout [17, 19]. DCE-MRI is not widely available, and different groups have proposed modified combinations of DCE and DWI for the evaluation of salivary neoplasms [17, 19, 24, 26].

Intravoxel Incoherent Motion Magnetic Resonance Imaging

The diffusion properties of neoplasms rely on cell density, whereas perfusion imaging offers physiologic tissue information. Using intravoxel incoherent motion (IVIM) MRI, the molecular diffusion and motion of water molecules in the capillary network can be estimated simultaneously [27–29]. Thus, a single DWI study with multiple b values can characterize the diffusion and perfusion properties of tumors without the use of intravenous contrast and appears to have potential in the differentiation of benign from malignant neoplasms [27–29]. IVIM characteristics were able to distinguish PAs, WTs and malignant neoplasms in 31 patients using a stepwise approach [29]. Larger studies are needed to define the role of IVIM in the head and neck [28].

Magnetic Resonance Spectroscopy

King et al. [30] applied in vivo proton magnetic resonance spectroscopy to characterize SG tumors. Choline (Cho) was detectable in all benign and malignant tumors, but not in normal parotid glands. However, the creatine (Cr) peak amplitude could be obtained in less than half of the neoplasms. Despite these limitations, WTs demonstrated the highest Cho/Cr ratio, and a Cho/Cr ratio of 2.4 might be used to differentiate benign from malignant lesions. These initial results have not been validated in a larger study, and they have not been reproduced.

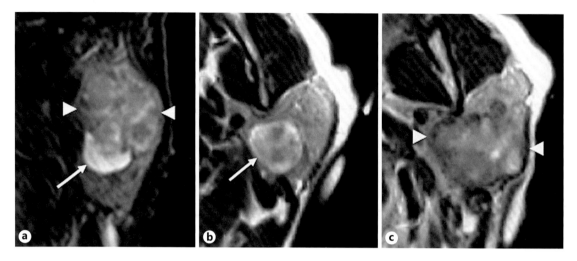

Fig. 6. Carcinoma ex-PA. Coronal short tau inversion recovery (**a**) and axial T2-weighted images (**b, c**) demonstrate a mass with a T2 hyperintense, well-defined component (the PA; arrows) inferiorly and a larger, more heterogeneous and T2 iso- to hypointense component superiorly (**c**), with irregular margins (the carcinoma; arrowheads).

Computerized Tomography Perfusion

A pilot study showed that CT perfusion was feasible for the evaluation of parotid neoplasms, with potential to differentiate benign and malignant tumors by demonstrating higher blood flow and volume in benign neoplasms, which was attributed to abundant microscopic areas of necrosis in malignancies [31]. A recent study was able to construct time-density curves for parotid neoplasms that appear similar to the time-intensity curves obtained with DCE-MRI [32]. Compared with malignant tumors, WTs had significantly higher blood flow and blood volume (sensitivity and specificity of 75 and 80%, respectively), and PAs had significantly lower blood flow and blood volume (sensitivity and specificity of 75 and 71%, respectively) [32]. Further studies are needed to justify the radiation exposure, which is higher than in conventional CT.

Positron Emission Tomography

Benign neoplasms with a high number of mitochondria, such as WTs, PAs and oncocytomas, will take up fluorodeoxyglucose (FDG), a glucose analog, which is the most commonly used tracer for PET imaging. Thus, PET cannot reliably differentiate these tumors from malignant tumors [33]. In fact, incidental focal parotid uptake is noted in approximately 10% of PET studies, due mostly to benign tumors (fig. 5) [34]. Conversely, some malignant tumors, such as adenoid cystic carcinoma, have variable or low uptake of FDG, and normal physiologic uptake by the SG can obscure these neoplasms [9]. Although not commonly used in practice, the combination of SG scintigraphy (to rule in or out a WT) and PET has a reported sensitivity and specificity for parotid malignancy of 75 and 80%, respectively [35]. PET/CT can detect nodal and distant metastatic disease, with potential impact on treatment planning, especially for high-grade malignancies [8, 9, 36, 37]. Following treatment, both false-positive and false-negative results can occur, and the accuracy of PET is not superior to that of MRI and/or CT [8, 9].

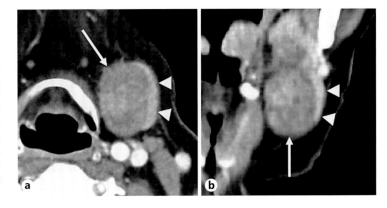

Fig. 7. Submandibular gland PA. Axial (**a**) and coronal (**b**) contrast-enhanced CT images demonstrate crescent-shaped compression of the submandibular gland (arrowheads) by a hypodense mass (arrows), which is a specific but insensitive sign of submandibular gland PA, secondary to compression of the normal gland parenchyma by the slow-growing tumor.

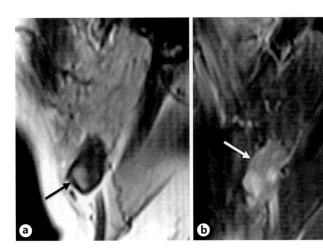

Fig. 8. Parotid tail WT. Note the T1 hyperintense cystic component (black arrow in **a**) in this enhancing (white arrow in **b**) parotid tail mass.

Imaging of Selected Salivary Gland Masses

PA, also known as a benign mixed tumor, is the most common SG neoplasm, and it can undergo malignant transformation, which will present with a lower T2 signal and diffusivity and more irregular margins than a benign adenoma (fig. 6) [38]. T2 hyperintensity, well-circumscribed borders and homogeneous contrast enhancement have a high positive predictive value (95%) in predicting PA when used in conjunction with FNA biopsy [39]. However, larger tumors can appear heterogeneous and demonstrate necrosis and hemorrhage. Calcifications favor PA diagnosis but are only occasionally present [2, 40]. In submandibular gland PA, crescent-shaped compression of the normal gland has been reported as a specific but insensitive sign (fig. 7) [41]. PAs are characterized by high ADC values and progressive enhancement (fig. 1) [17, 19, 24, 26].

WTs are the second most common parotid neoplasm, with less than 1% risk for malignant transformation. These tumors are frequently multiple and bilateral and tend to occur in the tail of the parotid gland (fig. 8) [2, 42]. Cystic components are seen in 30–50% of WTs and are usually hyperintense on T1, which has been suggested as a differentiating feature, especially relative to lymphoma (fig. 8) [4, 20]. Enhancement depends on the timing of the scan since these neoplasms dem-

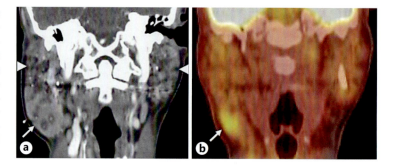

Fig. 9. MALT lymphoma in Sjögren's syndrome. An enhancing mass (arrows) with cystic changes (**a**) and increased metabolic activity on PET (**b**) arises from the inferior aspect of the right parotid gland. Note the characteristic changes of Sjögren's syndrome in the parotid glands (arrowheads in **a**), including lobules of glandular tissue with intervening fat and calcifications.

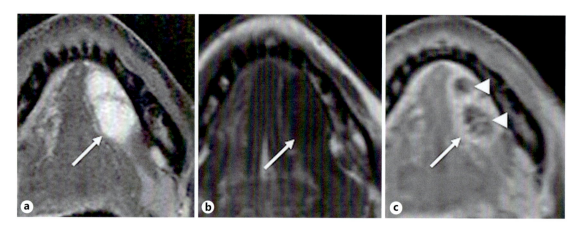

Fig. 10. Adenoid cystic carcinoma of the left sublingual gland. A well-defined mass (arrows) is centered in the left sublingual gland. The mass is hyperintense on T2 (**a**), which is usually a sign of a benign or low-grade malignant neoplasm. It is hypointense on T1 (**b**) and enhances, with central nonenhancing necrotic components (arrowheads in **c**).

onstrate early enhancement and high washout [17, 19, 23, 24]. WTs are hypermetabolic on PET (fig. 5) and accumulate Tc-99m, which can be used to differentiate them from other multiple masses, such as metastatic disease, lymphoma and sarcoidosis [2, 42]. An MRI-based decision algorithm using morphological, signal and functional data (DWI and DCE-MRI) achieved 80–85% sensitivity and 96–100% specificity for the presurgical diagnosis of WTs, which might be sufficient to avoid surgery when typical findings are present [43].

Oncocytoma is a rare SG neoplasm most commonly arising as a well-circumscribed mass in the parotid gland. Oncocytomas can appear hypermetabolic on PET and accumulate Tc-99m, which has been attributed to their high mitochondrial content. Isointensity to the native parotid gland on fat-suppressed T2W images and on postcontrast T1W images has been reported as a specific MRI finding of oncocytomas [44].

There is a much greater variety of malignant SG tumors, and their imaging appearance depends on their histologic type and grade. For example, low-grade mucoepidermoid carcinoma and acinic cell carcinoma can appear identical to benign tumors on conventional imaging and can

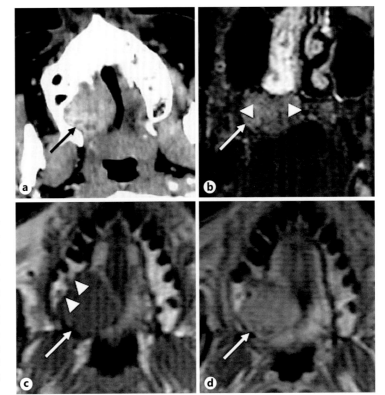

Fig. 11. Mucoepidermoid carcinoma of the hard palate. Contrast-enhanced CT (**a**) demonstrates an enhancing, well-circumscribed mass (black arrow) on the right side of the palate. On coronal T2 (**b**), the mass is hypointense (arrow) and erodes the hard palate (arrowheads). The mass (arrow) is hypointense on T1 (**c**) and scallops the adjacent maxillary alveolar ridge (arrowheads). The mass (arrow) enhances homogeneously following contrast administration (**d**).

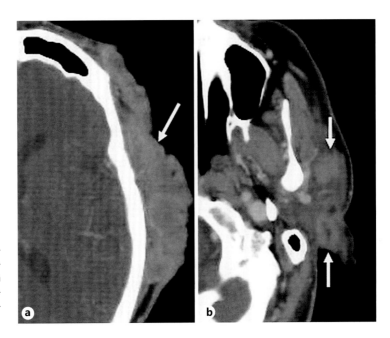

Fig. 12. Parotid lymph node metastases. Large scalp exophytic and enhancing squamous cell carcinoma (arrow in **a**) is associated with metastatic parotid lymphadenopathy (arrows in **b**).

even have similar ADC values, DCE-MRI features and FDG uptake as benign tumors, making diagnosis challenging (fig. 2). High-grade lesions are more likely to demonstrate ill-defined margins, T2 hypointensity, decreased ADC values and increased FDG uptake, and the goal of imaging is to determine local, perineural and metastatic spread (fig. 3, 4).

Primary SG lymphoma is rare. Mucosa-associated lymphoid tissue (MALT) lymphoma has a 44× higher incidence in patients with Sjögren's syndrome [2]. Therefore, any mass in a patient with Sjögren's syndrome should undergo tissue diagnosis. Other conditions associated with MALT lymphoma include rheumatoid arthritis and immunosuppression [45]. MALT lymphoma is often bilateral, is associated with cyst formation (fig. 9; hypointense on T1 in contrast to WTs), and demonstrates markedly decreased ADC values, similar to WTs.

Neoplasms of the submandibular and sublingual glands demonstrate similar signal characteristics to parotid neoplasms; however, a tumor in these glands is more likely to be malignant (fig. 10). The MSGs are dispersed within the submucosa of the upper aerodigestive tract, and they have the highest concentration within the buccal, labial, palatal, and lingual regions, which might explain the frequent involvement of the palate by MSG neoplasms (fig. 11) [12]. The MSG origin of these tumors is often not apparent immediately, and other mucosal or submucosal tumors are frequently considered in the differential diagnosis. The imaging appearance of an MSG malignancy, such as adenoid cystic carcinoma or mucoepidermoid carcinoma, is variable, depending on its grade, and might be indistinguishable from a mucosal squamous cell carcinoma.

Intraparotid Lymph Nodes
The parotid gland encapsulates late in the second trimester, allowing for incorporation of lymphoid tissue [42]. An intraparotid lymph node (IPLN) can have a lobulated appearance, and in conjunction with its T2 hyperintense signal, it might simulate a small PA or other tumor. Although not always present, identification of a fatty hilum is suggestive of an IPLN [12]. When available, ADC values allow differentiation of PA from an IPLN. IPLNs drain the anterior face, the lateral scalp, and the external auditory meatus and can become involved in skin malignancies, mainly squamous cell carcinoma and melanoma (fig. 12) [45]. Primary squamous carcinoma of the parotid gland can occur in the setting of chronic inflammation and squamous metaplasia; however, metastatic squamous carcinoma is much more common, usually presenting as a necrotic solid mass [12]. A solitary metastasis in the parotid gland will look indistinguishable from a high-grade neoplasm, but the usual presentation of metastatic disease is that of multiple, invasive masses, often with a necrotic component [12].

Conclusion

MRI is the method of choice in detecting, characterizing, staging and postoperatively following salivary neoplasms. Multiple advanced imaging techniques are under investigation, and DWI and DCE-MRI appear to improve diagnostic accuracy.

References

1 Friedman ER, Saindane AM: Pitfalls in the staging of cancer of the major salivary gland neoplasms. Neuroimaging Clin N Am 2013;23:107–122.
2 Shah GV: MR imaging of salivary glands. Neuroimaging Clin N Am 2004; 14:777–808.
3 Christe A, Waldherr C, Hallett R, et al: MR imaging of parotid tumors: typical lesion characteristics in MR imaging improve discrimination between benign and malignant disease. AJNR Am J Neuroradiol 2011;32:1202–1207.

4 Kato H, Kanematsu M, Watanabe H, et al: Salivary gland tumors of the parotid gland: CT and MR imaging findings with emphasis on intratumoral cystic components. Neuroradiology 2014;56: 789–795.

5 Ishibashi M, Fujii S, Kawamoto K, et al: Capsule of parotid gland tumor: evaluation by 3.0 T magnetic resonance imaging using surface coils. Acta Radiologica 2010;51:1103–1110.

6 Sakamoto M, Iikubo M, Kojima I, et al: Diagnostic value of capsule-like rim enhancement on magnetic resonance imaging for distinguishing malignant from benign parotid tumours. Int J Oral Maxillofac Surg 2014;43:1035–1041.

7 Moonis G, Cunnane MB, Emerick K, et al: Patterns of perineural tumor spread in head and neck cancer. Magn Reson Imaging Clin N Am 2012;20:435–446.

8 Park HL, Yoo Ie R, Lee N, et al: The Value of F-18 FDG PET for planning treatment and detecting recurrence in malignant salivary gland tumors: comparison with conventional imaging studies. Nucl Med Mol Imaging 2013; 47:242–248.

9 Roh JL, Ryu CH, Choi SH, et al: Clinical utility of 18F-FDG PET for patients with salivary gland malignancies. J Nucl Med 2007;48:240–246.

10 Ginsberg LE: Imaging pitfalls in the postoperative head and neck. Semin Ultrasound CT MR 2002;23:444–459.

11 Moonis G, Patel P, Koshkareva Y, et al: Imaging characteristics of recurrent pleomorphic adenoma of the parotid gland. AJNR Am J Neuroradiol 2007; 28:1532–1536.

12 Som PM, Brandwein MS: Anatomy and pathology of the salivary glands. Head Neck Imaging 2011;2:2448–2609.

13 Blitz AM, Choudhri AF, Chonka ZD, et al: Anatomic considerations, nomenclature, and advanced cross-sectional imaging techniques for visualization of the cranial nerve segments by MR imaging. Neuroimaging Clin N Am 2014; 24:1–15.

14 Thoeny HC: Diffusion-weighted MRI in head and neck radiology: applications in oncology. Cancer Imaging 2011;10: 209–214.

15 Thoeny HC, De Keyzer F, King AD: Diffusion-weighted MR imaging in the head and neck. Radiology 2012;263: 19–32.

16 Ikeda M, Motoori K, Hanazawa T, et al: Warthin tumor of the parotid gland: diagnostic value of MR imaging with histopathologic correlation. AJNR Am J Neuroradiol 2004;25:1256–1262.

17 Yabuuchi H, Matsuo Y, Kamitani T, et al: Parotid gland tumors: can addition of diffusion-weighted MR imaging to dynamic contrast-enhanced MR imaging improve diagnostic accuracy in characterization? Radiology 2008;249: 909–916.

18 Habermann CR, Arndt C, Graessner J, et al: Diffusion-weighted echo-planar MR imaging of primary parotid gland tumors: is a prediction of different histologic subtypes possible? AJNR Am J Neuroradiol 2009;30:591–596.

19 Eida S, Sumi M, Nakamura T: Multiparametric magnetic resonance imaging for the differentiation between benign and malignant salivary gland tumors. J Magn Reson Imaging 2010;31:673–679.

20 Kato H, Kanematsu M, Goto H, et al: Mucosa-associated lymphoid tissue lymphoma of the salivary glands: MR imaging findings including diffusion-weighted imaging. Eur J Radiol 2012; 81:e612–e617.

21 Kolff-Gart AS, Pouwels PJ, Noij DP, et al: Diffusion-weighted imaging of the head and neck in healthy subjects: reproducibility of ADC values in different MRI systems and repeat sessions. AJNR Am J Neuroradiol 2015;36:384–390.

22 Hwang I, Choi SH, Kim YJ, et al: Differentiation of recurrent tumor and post-treatment changes in head and neck squamous cell carcinoma: application of high b-value diffusion-weighted imaging. AJNR Am J Neuroradiol 2013;34: 2343–2348.

23 Yabuuchi H, Fukuya T, Tajima T, et al: Salivary gland tumors: diagnostic value of gadolinium-enhanced dynamic MR imaging with histopathologic correlation. Radiology 2003;226:345–354.

24 Lam PD, Kuribayashi A, Imaizumi A, et al: Differentiating benign and malignant salivary gland tumors: diagnostic criteria and the accuracy of dynamic contrast-enhanced MRI with high temporal resolution. Br J Radiol 2015: 20140685.

25 Asaumi J, Yanagi Y, Hisatomi M, et al: The value of dynamic contrast-enhanced MRI in diagnosis of malignant lymphoma of the head and neck. Eur J Radiol 2003;48:183–187.

26 Espinoza S, Malinvaud D, Siauve N, et al: Perfusion in ENT imaging. Diagn Interv Imaging 2013;94:1225–1240.

27 Sumi M, Nakamura T: Head and neck tumors: assessment of perfusion-related parameters and diffusion coefficients based on the intravoxel incoherent motion model. AJNR Am J Neuroradiol 2013;34:410–416.

28 Sumi M, Nakamura T: Head and neck tumours: combined MRI assessment based on IVIM and TIC analyses for the differentiation of tumors of different histological types. Eur Radiol 2014;24: 223–231.

29 Sumi M, Van Cauteren M, Sumi T, et al: Salivary gland tumors: use of intravoxel incoherent motion MR imaging for assessment of diffusion and perfusion for the differentiation of benign from malignant tumors. Radiology 2012;263:770–777.

30 King AD, Yeung DK, Ahuja AT, et al: Salivary gland tumors at in vivo proton MR spectroscopy. Radiology 2005;237: 563–569.

31 Bisdas S, Baghi M, Wagenblast J, et al: Differentiation of benign and malignant parotid tumors using deconvolution-based perfusion CT imaging: feasibility of the method and initial results. Eur J Radiol 2007;64:258–265.

32 Dong Y, Lei GW, Wang SW, et al: Diagnostic value of CT perfusion imaging for parotid neoplasms. Dentomaxillofac Radiol 2014;43:20130237.

33 Toriihara A, Nakamura S, Kubota K, et al: Can dual-time-point 18F-FDG PET/CT differentiate malignant salivary gland tumors from benign tumors? AJR Am J Roentgenol 2013;201:639–644.

34 Treglia G, Bertagna F, Sadeghi R, et al: Prevalence and risk of malignancy of focal incidental uptake detected by fluorine-18-fluorodeoxyglucose positron emission tomography in the parotid gland: a meta-analysis. Eur Arch Otorhinolaryngol 2014;1–10.

35 Uchida Y, Minoshima S, Kawata T, et al: Diagnostic value of FDG PET and salivary gland scintigraphy for parotid tumors. Clin Nucl Med 2005;30:170–176.

36 Kim MJ, Kim JS, Roh JL, et al: Utility of 18F-FDG PET/CT for detecting neck metastasis in patients with salivary gland carcinomas: preoperative planning for necessity and extent of neck dissection. Ann Surg Oncol 2013;20: 899–905.

37 Razfar A, Heron DE, Branstetter BF, et al: Positron emission tomography-computed tomography adds to the management of salivary gland malignancies. Laryngoscope 2010;120:734–738.

38 Kato H, Kanematsu M, Mizuta K, et al: Carcinoma ex pleomorphic adenoma of the parotid gland: radiologic-pathologic correlation with MR imaging including diffusion-weighted imaging. AJNR Am J Neuroradiol 2008;29:865–867.

39 Heaton CM, Chazen JL, van Zante A, et al: Pleomorphic adenoma of the major salivary glands: diagnostic utility of FNAB and MRI. Laryngoscope 2013; 123:3056–3060.

40 Abdullah A, Rivas FF, Srinivasan A: Imaging of the salivary glands. Semin Roentgenol 2013;48:65–74.

41 Kashiwagi N, Murakami T, Nakanishi K, et al: Conventional MRI findings for predicting submandibular pleomorphic adenoma. Acta Radiologica 2013;54: 511–515.

42 Yousem DM, Kraut MA, Chalian AA: Major salivary gland imaging. Radiology 2000;216:19–29.

43 Espinoza S, Felter A, Malinvaud D, et al: Warthin's tumor of parotid gland: surgery or follow-up? Diagnostic value of a decisional algorithm with functional MRI. Diagn Interv Imaging 2016;97: 37–43.

44 Patel ND, van Zante A, Eisele DW, et al: Oncocytoma: the vanishing parotid mass. AJNR Am J Neuroradiol 2011;32: 1703–1706.

45 Lee YY, Wong KT, King AD, et al: Imaging of salivary gland tumours. Eur J Radiol 2008;66:419–436.

Marinos Kontzialis
Division of Neuroradiology, Russell H. Morgan Department of Radiology and Radiological Science
The Johns Hopkins University School of Medicine
600 North Wolfe Street
Baltimore, MD 21287 (USA)
E-Mail marinos.kontzialis@gmail.com

Evaluation: Fine Needle Aspiration Cytology, Ultrasound-Guided Core Biopsy and Open Biopsy Techniques

David C. Howlett[a] · Asterios Triantafyllou[b]

[a]Eastbourne Hospital, Eastbourne, and [b]Oral and Maxillofacial Pathology, School of Dentistry, University of Liverpool, Liverpool, UK

Abstract

The optimum technique for biopsy assessment of the nature of a major salivary gland mass remains controversial. Fine needle aspiration cytology (FNAC) has been the traditional and popular choice, but sampling of cellular clusters is largely associated with high non-diagnostic and false-negative rates, even under optimised circumstances. Ultrasound-guided core biopsy (USCB) provides a core of tissue that allows preservation of tissue architecture and that can be histologically and immunohistochemically examined, thereby improving the chances of a meaningful diagnosis. Although relatively recently applied in the pre-operative investigation of salivary lesions, USCB shows higher levels of accuracy and reduced non-diagnostic rates when compared with FNAC, in addition to good patient tolerability. A degree of caution should, however, be exercised because of the potential for tumour seeding, and time delays inherent to histological processing are also unavoidable. Where available, USCB may be given preference as the biopsy technique of choice in major salivary gland diagnosis. In units where FNAC performs well, USCB can be utilised when FNAC is non-diagnostic or equivocal. Intra-operative frozen section collection is invasive but may offer a secondary option in cases of non-diagnostic FNAC and/or USCB or when USCB is not available.

© 2016 S. Karger AG, Basel

Introduction

It is good practice that patients presenting with a palpable swelling of the parotid or submandibular salivary glands should be investigated in an attempt to obtain an accurate diagnosis prior to consideration for surgery. The pre-operative diagnostic process usually involves initial clinical evaluation, imaging (generally ultrasound in the first instance) and then biopsy with assessment of pathology where appropriate. Investigation should occur in a manner that is both timely and standardised. The advantages of a pre-operative diagnosis are manifold. Pre-operative assessment of pathology would allow triage of patients in terms of timing and type of operative intervention; consideration of pre-operative adjuvant/

neo-adjuvant therapy, if indicated; and avoidance of surgery in select patients, especially for some tumour types where observation is appropriate (e.g. Warthin tumour) and for lymphoma, for which treatment is non-surgical. A definitive pre-operative diagnosis also facilitates informed patient consent in surgical cases, with particular emphasis on the need to possibly resect the facial nerve. Lastly, appreciation of lesion pathology is required when minimally invasive surgical techniques are being considered (e.g. extra-capsular dissection).

Biopsy Assessment of Pathology Pre-Operatively

This chapter focuses on the third and final component of the pre-operative diagnostic process, namely, biopsy assessment of pathology. The optimum biopsy technique for salivary gland diagnosis remains controversial. Here, we discuss the two main options in current clinical practice: fine needle aspiration cytology (FNAC) and fine needle core biopsy, usually under ultrasound guidance (ultrasound-guided core biopsy, USCB). The potential role of intra-operative frozen sections (IOFSs) is also addressed.

Traditionally, many major salivary gland lesions were treated by surgical excision, which acted as both a diagnostic and a therapeutic manoeuvre. Inevitably, some lesions were inadequately or incompletely excised, or lesions that would have been better managed non-surgically were removed. To address these problems, the technique of open surgical biopsy was developed, which would then guide further intervention. This, however, fell out of favour due to associated complications, which included facial nerve injury, infection, fistula, sialocele formation and tumour spillage and recurrence [1]. By the early 1980s, open biopsy had been largely superseded by FNAC.

The Use of Fine Needle Aspiration Cytology

FNAC was originally performed with guidance by palpation, without image guidance, by clinicians in the outpatient clinic setting. The technique is well described and widely endorsed because of many advantages: it is quick, safe, relatively non-invasive and inexpensive. There have been numerous publications on the diagnostic efficacy of FNAC, and its application in salivary gland assessment has been regarded as successful. Nevertheless, it has become increasingly apparent that FNAC without image guidance may be associated with high non-diagnostic rates, high false-positive and false-negative rates and relatively poor sensitivity/specificity [2, 3]. Other approaches have thus been explored to improve FNAC performance. Ultrasound guidance of FNAC (as with core biopsy) allows direct needle visualisation and placement, along with accurate sampling of solid tumour foci in complex (variously solid and cystic) lesions as well as avoidance of adjacent vascular structures [4]. Performance of FNAC can be further improved if a cytopathologist or cytology technician attends the clinical session to assess the sample's adequacy [5]. Other clinical diagnostic models present the cytopathologist with the opportunity for direct assessment of aspirates and repeat sampling, as necessary; ancillary cytology technology (in situ hybridisation, flow cytometry) may also be considered in larger units [4, 6], though these techniques are time consuming and preclude repeat sampling.

Reliance on Fine Needle Aspiration Cytology Alone

Despite these optimisation manoeuvres for FNAC, concerns persist, as discussed in a large meta-analysis published in 2011 [7]. This analysis reviewed findings from 64 studies published since the late 1980s and included FNAC data collected not only blindly but also under varying levels of optimisation, as already discussed. Overall, FNAC was found to be safe and well tolerated, as expected, with high reported specificity (97%) but a lower sensitivity (80%). A positive diagnosis was reliable, particularly in benign disease, which may reflect the increased occurrence of the confidently diagnosed pleomorphic adenoma and Warthin tumour in the parotid glands, but there was a high false-negative rate (20%). It is also notable that although not all studies included non-diagnostic rates, where available, the figure for non-diagnostic sampling was 8%. The review also showed a statistically significant wide variation in performance across centres. It is likely that salivary FNAC is often reported by cytopathologists or general pathologists with a variable interest in the head and neck, whereas specialists in salivary gland histopathology often show little or no interest in cytopathology and argue that the procedure induces glandular responses, causing diagnostic difficulties during the final histological examination. Post-FNAC glandular responses (see [8] for a review) resemble those following salivary infarction and include squamous metaplasia that is occasionally prominent, necrosis, 'ghost' architecture, inflammation and a myofibroblastic reaction. It is also likely that FNAC unsatisfactorily assesses 'double' salivary pathology (e.g. a Warthin tumour and IgG4-related sclerosing disease). Although the difficulties have been overemphasised here, it is important to appreciate that a precise pre-operative diagnosis, although desirable, is not always achievable.

FNAC performance is likely to be enhanced in a specialised unit with experienced operators and an ultrasound-guided and/or cytopathologist/pathologist-led clinical diagnostic model. A recent publication reported a 6% false-negative rate and a 10% false-positive rate (out of 138 cases), a sensitivity for malignancy of 73% and a specificity of 87% in a cytologist-led clinic [9]. However, the necessary optimised circumstances are not currently widely available, and in particular, cytopathologists, cytology technicians and ancillary cytology facilities are in short supply outside of larger institutions. It is now increasingly recognised that salivary gland lesions may be better managed in dedicated, specialised units.

The Need for More than Cells for Diagnosis

The well-recognised diagnostic difficulties for FNAC reflect the inherent disadvantages of the various cytological modalities. Even when an adequate aspirate is obtained, FNAC provides variously clustered cells, and not tissue with architecturally arranged cells and supporting stroma. It is this architectural preservation in a biopsy sample that will allow accurate diagnosis, and many of the more common salivary tumours show similar cellular phenotypes [10]. Examples include cellular pleomorphic adenoma, monomorphic adenoma, basal cell adenoma and adenocarcinoma, adenoid cystic carcinoma and low-grade mucoepidermoid carcinoma [9].

An Alternative: Ultrasound-Guided Core Biopsy

The persisting debate around FNAC has led interested clinicians to explore alternative approaches, and in particular USCB. USCB was initially established in breast and abdominal diagnosis and was first successfully applied in a series of parotid patients in 1999 [11]. There have been several subsequent published series, 12 of which were included in a recent meta-analysis [12]. The technique of USCB is also well described [13, 14]. It requires an operator trained in diagnostic head and neck ultrasound and biopsy techniques and involves infiltration of local anaesthetic and introduction of a small-core biopsy needle under ultrasound guidance via a small skin incision. A variety of automated, spring-loaded biopsy devices exist. The needle throw can often be adjusted, and once deployed, the needle obtains a core

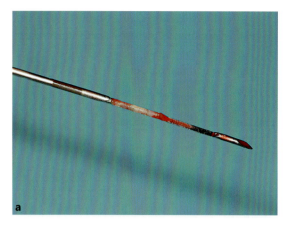

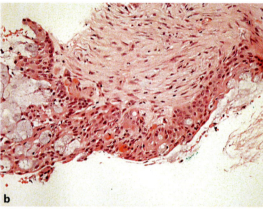

Fig. 1. An image of a typical core sample obtained using an 18G needle measuring 1.2 mm thick. The core sample can be seen within the biopsy tray, revealed by retracting the overlying needle sheath. **b** Haematoxylin- and eosin-stained histological section of the sample from **a** shows a non-high-grade mucoepidermoid carcinoma. Note mucous cells (left of the image) and squamoid epithelium (right of the image). Stroma is also present (top of the image).

of tissue with a preserved architectural arrangement and relationship of cells and stroma. The size of the needle and core can be varied, but most parotid biopsy series utilise 18- or 20-gauge (G) needles, usually with 1–3 needle passes. The core(s) of tissue obtained are routinely processed for histology and can be further subjected to immunohistochemical analysis, thus significantly increasing the chances of achieving a meaningful diagnosis (e.g. differentiating between undifferentiated carcinoma and lymphoma, recognising neuroendocrine carcinoma) (fig. 1).

Grading may also be attempted, but there are limitations, particularly in lymphomas (even salivary mucosa-associated lymphoid tissue lymphomas), in which case a service organisation may request reporting by haematolymphoid, rather than head and neck, pathologists.

The Evidence for Ultrasound-Guided Core Biopsy

The published meta-analysis [12] acknowledged a smaller number of publications on USCB over a shorter timescale than similar FNAC meta-analyses have. The results were, however, considered valid and useful for future comparisons as regards parotid USCB. When compared with FNAC, USCB showed increased sensitivity (96%) and specificity (100%) and a decreased non-diagnostic rate (1.6%). The good safety profile and patient tolerability of USCB were strengthened, with only 8 haematomas reported (1.6%). In addition, the procedure was undertaken in the outpatient setting and did not show variability in performance across participating centres, which is a feature of FNAC. The latter, however, may merely reflect the established diagnostic superiority of tissue *vs.* cellular aspirate. A consensus is easier to achieve when tissue is assessed by pathologists with expertise in salivary histopathology, rather than when aspirates are assessed by cytologists with an interest in the head and neck.

Controversies over the Use of Ultrasound-Guided Core Biopsy

Despite the advantages noted above, as a technique, USCB remains controversial. USCB is more invasive than FNAC, and theoretically, there is a risk of vascular or parotid facial nerve damage, although these complications are not described and should be readily avoidable when USCB is supplemented by using ultrasound guidance. USCB also does not readily lend itself to a 'one-stop' clinical model, which is more suited to FNAC, as there is an inbuilt delay around processing and reporting of USCB specimens, which is common to all procedures (biopsies, resections) where tissue, rather than cells, is obtained. FNAC-based one-stop clinics seem attractive, but published work suggests that alternative diagnostic clinics including USCB as the biopsy tool of choice can offer equally or more effective patient care [15].

Displaced epithelia and tumour seeding are potential complications of USCB that need particular consideration in the case of salivary gland neoplasms, and particularly pleomorphic adenoma, which has recognised potential for tumour seeding when violated. There were no reports of tumour seeding in the published meta-analysis [12]. A comparative evaluation published in 2013, which addressed tumour seeding post-salivary gland biopsy [16], found only two cases of tumour seeding after a 14G core needle biopsy, with two cases also described post-FNAC. As tumour seeding post-salivary biopsy may occur up to 20 years following biopsy and as the follow-up periods for USCB studies are relatively short (up to 16 years in the most recent published series [17]), it is clear that continued surveillance is needed. Tumour seeding probably relates to needle size and is rare with the 18/20G needles usually used in salivary gland biopsy. The option is available for surgeons to excise the biopsy tract at the time of surgery, although there is no evidence to suggest that this approach is routinely required.

The article alluded to in the paragraph above [17] contains data from 313 parotid neoplasms that underwent USCB, serving as the largest currently published series, but this study was not included in the 2014 meta-analysis [12]. In comparison with the meta-analysis, the article reported a higher non-diagnostic rate, or 4%, which is thought to relate to the difficulty obtaining suffi-

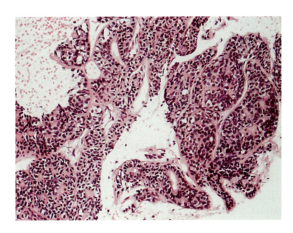

Fig. 2. A biphasic architectural pattern with abluminal clear cells is recognised. The appearances are shared by various epithelial salivary tumours. The definitive diagnosis of basal cell adenoma was achieved only when the resected tumour became available for histological examination (haematoxylin and eosin staining).

cient material from highly cystic lesions, a problem shared by FNAC. It also reported two false-negative results: USCB results were reported as basal cell adenoma and benign myoepithelioma, but histological examination of the surgical resections, enabling assessment of the tumour silhouette/interface along with adjacent tissues, established the correct diagnoses, or basal cell adenocarcinoma and myoepithelial carcinoma, respectively. Assessment of the margin/relationship of a lesion with surrounding soft tissues, including the presence of a capsule or patterns of invasion, is important in reaching a definitive diagnosis. A part of the margin may be included in USCB, but not reliably, which could be a significant cause of diagnostic failure in some cases, despite provision of a core of tissue.

Another pitfall for USCB results from the limited width of the core. Hence, the features of the tissue obtained may be not representative or sufficiently characteristic for diagnosis or may be highly unusual (fig. 2).

This pitfall is not unique to USCB, but rather reflects the problem of obtaining an adequate volume of tissue for a histological diagnosis and is a common experience in conventional, incisional biopsies of mucosae or skin. As is the case with FNAC, although less commonly, an accurate preoperative diagnosis via USCB is not always feasible.

Intra-Operative Frozen Sections

The final salivary gland biopsy technique to be discussed is that of IOFSs, which have experienced a minor resurgence as another alternative to the underperformance of FNAC. IOFSs were the subject of a meta-analysis published in 2011, which contained data from 13 studies over the 1985–2010 period [18]; a 2014 publication by Fakhry et al. [9] is an additional relevant reference. While the former meta-analysis described 90% sensitivity, 99% specificity and consistency of results across study centres [18], the latter reported a sensitivity of 80% and a specificity of 98% [9]. There are clinical concerns about IOFSs, as these share many of the potential problems previously described for open surgical biopsy. In addition, many pathologists, and particularly those working outside of specialist units, are unwilling to provide a definitive diagnosis in what is a challenging diagnostic area in the IOFS scenario, and the pitfalls/failures already outlined for USCB also apply to IOFSs.

It has been suggested that IOFSs can complement non-diagnostic or equivocal FNAC since a pre-operative diagnosis is desired by both clinicians and patients for the reasons discussed earlier, so efforts should be directed at achieving this. There are even situations where surgical diagnostic excisional biopsy may be necessary, including parotid deep lobe lesions not amenable to percutaneous biopsy and also lesions that are suspicious on clinical/imaging assessment and where FNAC and USCB are benign or considered non-representative (fig. 2).

Conclusion

The optimum means for biopsy assessment of a salivary gland lesion is evolving and remains controversial. Overall, the situation reflects the perennial problems of cytology *vs.* histology and which technique can ensure the volume of tissue adequate for diagnosis. Where available, USCB as a stand-alone technique seems the best in terms of diagnostic adequacy and accuracy, although some concerns persist around the potential for tumour seeding. Longer follow-up of patients who have had USCB is needed to address this controversy. It is acknowledged that FNAC can perform effectively under optimised circumstances, but diagnostic issues remain inherent to a technique reliant on provision of a cellular aspirate, without preservation of tissue architecture. In institutions where FNAC is working well, USCB should be considered the second-line biopsy modality of choice when FNAC is non-diagnostic or equivocal; IOFSs may be reserved for centres that do not have access to USCB.

References

1. McGuirt WF, McCabe BF: Significance of node biopsy before definitive treatment of cervical metastatic carcinoma. Laryngoscope 1978;88:594–597.
2. Balakrishnan K, Castling B, McMahon J, et al: Fine needle aspiration cytology in the management of a parotid mass: a two centre retrospective study. Surgeon 2005;2:67–72.
3. Mallon DH, Kostalas M, MacPherson FJ, et al: The diagnostic value of fine needle aspiration of parotid lumps. Ann R Coll Surg Engl 2013;95:258–262.
4. Robinson IA, Cozens N: Does a joint ultrasound guided cytology clinic optimise the cytological evaluation of head and neck masses? Clin Radiol 1999;54:312–316.
5. Eisele DW, Sherman ME, Koch WM, et al: Utility of immediate on-site cytopathological procurement and evaluation in fine needle aspiration biopsy of head and neck masses. Laryngoscope 1992;102:1328–1330.
6. Coghill S, Brown S: Why pathologists should take needle aspiration specimens. Cytopathology 1991;2:67–74.

7 Schmidt RL, Hall BJ, Wilson AR, et al: A systematic review and meta-analysis of the diagnostic accuracy of fine needle aspiration cytology for parotid gland lesions. Am J Clin Pathol 2011;136:45–59.
8 Triantafyllou A, Hunt JL, Devaney KO, et al: A perspective of comparative salivary and breast pathology. Part I: microstructural aspects, adaptations and cellular events. Eur Arch Otorhinolaryngol 2014;271:647–663.
9 Fakhry N, Santini A, Lagier A, et al: Fine needle aspiration cytology and frozen section in the diagnosis of malignant parotid tumours. Int J Oral Maxillofac Surg 2014;43:802–805.
10 Nelson BL, Thompson LDR: Incisional or core biopsies of salivary gland tumours: how far should we go? Diagnostic Histopathology 2012;18:358–365.
11 Buckland JR, Manjaly G, Violaris N, et al: Ultrasound-guided cutting needle biopsy of the parotid gland. J Laryngol Otol 1999;113:988–992.
12 Witt BL, Schmidt RL: Ultrasound-guided core needle biopsy of salivary gland lesions: a systematic review and meta-analysis. Laryngoscope 2014;124:695–700.
13 Breeze J, Andi A, Williams MD, et al: The use of fine needle core biopsy under ultrasound guidance in the diagnosis of a parotid mass. Br J Oral Maxillofac Surg 2009;47:78–79.
14 Howlett DC, Menezes LJ, Lewis K, et al: Sonographically guided core biopsy of a parotid mass. Am J Roentgenol 2007;188:223–227.
15 Cozens NJ: A systematic review that evaluates one-stop neck lump clinics. Clin Otolaryngol 2009;34:6–11.
16 Douville NJ, Bradford CR: Comparison of ultrasound guided core biopsy versus fine needle aspiration biopsy in the evaluation of salivary gland lesions. Head Neck 2013;25:1657–1661.
17 Haldar S, Mandalia U, Skelton E, et al: Diagnostic investigation of parotid neoplasms – a 16-year experience of freehand fine needle aspiration cytology and ultrasound guided core biopsy. Int J Oral Maxillofac Surg 2015;44:151–157.
18 Schmidt RL, Hunt JP, Hall BJ, et al: A systematic review mcta-analysis of the diagnostic accuracy of frozen section for parotid gland lesions. Am J Clin Pathol 2011;136:729–738.

David C. Howlett, Consultant Radiologist and Honorary Clinical Professor
Eastbourne Hospital
King's Drive
Eastbourne BN21 2UH (UK)
E-Mail david.howlett@nhs.net

Facial Nerve Monitoring

Orlando Guntinas-Lichius[a] · David W. Eisele[b]

[a]Department of Otorhinolaryngology, Jena University Hospital, Jena, Germany; [b]Department of Otolaryngology – Head and Neck Surgery, The Johns Hopkins University School of Medicine, Baltimore, Md., USA

Abstract

Facial nerve monitoring has been increasingly routinely used as an intraoperative adjunctive method to help the head and neck surgeon to identify and minimize facial nerve injury during parotid surgery. The goals, current applications, recent technical advances, and limitations of the method are reviewed. A main focus of this chapter is a review of several prospective clinical trials that have been performed in recent years that have analyzed the benefit of electrophysiological nerve monitoring during parotid surgery. It has been demonstrated that nerve monitoring reduces the risk of early postoperative facial nerve dysfunction in primary surgery, but not in revision surgery. The effect is more pronounced in total than in superficial parotidectomy. Monitoring is associated with shorter surgical times in primary superficial parotidectomy compared to total parotidectomy. Facial nerve stimulation at the completion of parotidectomy helps to prognosticate the facial nerve functional outcome. A lower postdissection to predissection ratio of the maximal response amplitude is associated with early postoperative facial dysfunction. Facial nerve monitoring also helps the surgeon to avoid facial nerve injury when the facial nerve is not exposed during parotid surgery, such as during extracapsular dissection of a parotid neoplasm or sentinel node biopsy.

© 2016 S. Karger AG, Basel

Introduction

Facial nerve injury is the most serious complication of parotid surgery. Facial nerve paralysis can cause not only motor deficits but also cosmetic and functional morbidity; ocular complications; problems when eating or drinking; and, as a consequence, a diminished quality of life [1, 2]. Temporary facial nerve dysfunction occurs in 20–40% of patients undergoing parotidectomy, whereas permanent facial nerve dysfunction occurs in only 0–4% of patients [3–6]. Beyond discussions on the optimal surgical technique to preserve the facial nerve, facial nerve monitoring is the most discussed method to prevent immediate and permanent postoperative facial palsy in patients undergoing parotid surgery. This follows the results of the wide and efficient application of facial nerve monitoring in otologic, neurotologic, and skull base surgery [7, 8]. The routine use of facial nerve monitoring in vestibular schwannoma surgery has demonstrated improved preservation of facial nerve function [9, 10]. Its cost-effectiveness has also been shown for otologic surgery [11, 12]. About 75% of otolaryngologist-head and neck surgeons in Germany and over 67–80% in the

United Kingdom use nerve monitoring during parotid surgery [6, 13, 14]. A survey from 2005 demonstrated that 60% of otolaryngologist-head and neck surgeons in the United States use facial nerve monitoring during parotid surgery [15]. A higher percentage of surgeons (79%) who performed more than 10 parotidectomies per year used monitoring. Routine use during parotid surgery is more common among otolaryngologist-head and neck surgeons than among oral and maxillofacial surgeons in the United Kingdom [16].

Despite the high levels of usage, data on the efficacy of facial nerve monitoring in parotid surgery remain unclear. So far, there have been no randomized controlled trials performed to assess the efficacy of intraoperative facial nerve monitoring [17]. Retrospective studies are generally underpowered due to the overall low frequency of facial nerve injury during parotid surgery.

Ideally, facial nerve monitoring during parotid surgery would allow (1) early nerve identification, (2) warning of the surgeon of unexpected facial nerve stimulation, (3) mapping of the course of the nerve, (4) reduction of mechanical nerve trauma, and (5) evaluation and prognostication of function at the conclusion of the procedure.

Electrophysiology and the Basis of Facial Nerve Monitoring

The simplest approach to monitor the facial nerve is to visually monitor facial nerve movements during surgery (optical facial nerve monitoring). This approach, however, is not the focus of this chapter. Nevertheless, optical monitoring is often combined with electrophysiological monitoring in order to confirm the proper functioning of the electrophysiological monitoring equipment and the electrical nerve stimulator.

In general terms, facial nerve monitoring means electrophysiological monitoring. This technique provides for monitoring of facial muscle activity via electromyography (EMG). The EMG is typically recorded with needle electrodes placed in facial nerve muscles. Voluntary activation of the facial nerve in the alert patient as well as intraoperative electrical stimulation or mechanical irritation during parotid surgery generates motor evoked potentials on the basis of temporal and spatial summation of facial motor neuron excitatory postsynaptic potentials [18]. Therefore, neuromuscular blockade should be avoided for facial nerve monitoring [19].

Typically, facial muscle EMG is monitored and interpreted subjectively by an electrophysiologist or by the surgical team during parotid surgery. Signals are both visual and audible. Recently, automated analysis of A-train activity using intraoperative facial nerve EMG during vestibular schwannoma surgery has been established [20, 21]. This approach seems to allow for automatic calculation of the individual risk for postoperative facial palsy. Automated EMG monitoring analysis has not yet been established for parotid surgery.

Systems for Facial Nerve Monitoring

Several electrophysiological multichannel nerve-monitoring systems are commercially available. Relevant technical innovations in recent years have not been reported. The systems typically have 2–8 channels (fig. 1). In parotid surgery, most systems have 2 channels, and most data are published for 2-channel systems [17]. All systems continuously track facial muscle activity during surgery (passive monitoring; fig. 2) and have a built-in pulse generator for electrically evoked EMG responses (active monitoring; fig. 3). There are no data showing that systems with greater than 2 channels are more effective than 2-channel systems. Furthermore, it has not been demonstrated that a combination of passive and active monitoring is superior to passive monitoring

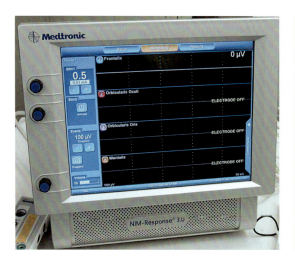

Fig. 1. Typical 4-channel EMG monitoring device with recording from the frontal muscle, orbicularis oculi muscle, orbicularis oris muscle, and mentalis muscle. The current of the stimulation electrode is set to 0.5 mA. The threshold to capture the EMG potential from the recorded muscle is set to 100 μV.

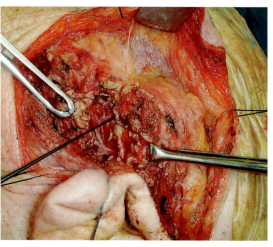

Fig. 3. Active stimulation of the facial nerve during parotidectomy. The facial nerve is exposed, and a peripheral branch is stimulated with the nerve stimulator.

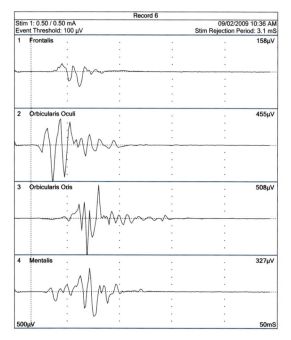

Fig. 2. Typical 4-channel EMG recording after stimulation of the main trunk of the facial nerve.

alone in protecting the facial nerve. The relationship of the results of active monitoring at the beginning and the completion of parotid surgery, however, might be useful to predict the facial nerve outcome [22].

Indications

Reasonable applications for facial nerve monitoring during parotid surgery include all instances in which facial nerve identification, dissection, and preservation are potentially difficult [23]. Such situations especially include reoperation, prior radiation therapy, malignant neoplasms, and large or deep lobe neoplasms that contribute to anatomic distortion, chronic parotitis, and minimally invasive surgical procedures such as intraparotid sentinel lymph node biopsy [24]. Similar to its role during sentinel lymph node biopsy, it is highly recommended to routinely conduct facial nerve monitoring when performing an extracapsular dissection of a parotid neoplasm [25]. By definition, extracapsular dissection is guided by the tumor, and not by anatomic dissection of the facial nerve. The main trunk of the facial nerve is not exposed during extracapsular dissection. Therefore, in facial nerve monitoring, it is useful to confirm the identity of the facial nerve by warning the surgeon, through mechanically evoked responses, when surgical dissection occurs in close proximity to the main trunk of the facial nerve or a facial nerve branch. Intraoperative facial nerve monitoring may also be used in the removal of selected intraparotid foreign bodies, without parotidectomy and dissection of the facial nerve, or during combined approaches to the removal of parotid sialoliths [26].

Benefits

Some surgeons use facial nerve monitoring routinely for all parotid surgery. The benefits of routine use of facial nerve monitoring include its intraoperative availability if an unanticipated need for its use arises. Routine use ensures surgeon familiarity with the nerve-monitoring system and facility with methods of troubleshooting the system [23]. After an initial learning period, the surgeon can properly interpret various signals and differentiate artifact from true events. Routine use of nerve monitoring may also result in a reduction in operative time [27].

Ideally, the benefits of facial nerve monitoring should be to assist in nerve identification and to give the surgeon confidence by warning the surgeon when mechanically evoked EMG responses occur due to the proximity of the facial nerve. In addition, an important benefit of active monitoring is differentiation of facial nerve branches from nonnerve tissues and sensory nerves. Active stimulation can also be used for mapping of the course of the facial nerve. All of these benefits should result in a reduction of mechanical nerve trauma and therefore postoperative facial palsy. Another desirable effect should be allowing for prognostication of postoperative facial function at the conclusion of the surgical procedure. Some of these factors have been addressed by recent prospective clinical trials as outcome parameters.

Sood et al. [17] performed a meta-analysis of the literature from 1970 to 2014 to determine the effectiveness of intraoperative facial nerve monitoring in preventing immediate and permanent postoperative facial nerve weakness in patients undergoing primary parotidectomy. Seven out of 1,414 articles met the inclusion criteria, including a total of 546 patients. The incidence of immediate postoperative weakness following parotidectomy was significantly lower in the monitored group compared to the unmonitored group (22.5 vs. 34.9%; $p = 0.001$). The incidence of permanent weakness was not significantly different in the long term (3.9 vs. 7.1%; $p = 0.18$). The number of monitored cases needed to prevent one case of immediate postoperative facial nerve weakness was 9, given an absolute risk reduction of 11.7% [17]. The authors also note that, to date, no randomized controlled trials have assessed the efficacy of facial nerve monitoring during parotid surgery.

Grosheva et al. [27] compared EMG monitoring combined with visual monitoring to visual monitoring alone during parotidectomy in a prospective trial with 50 patients in each group. EMG monitoring had no significant effect on the immediate postoperative or long-term facial function outcome ($p = 0.23$ and $p = 0.45$, respectively). The duration of superficial, but not total, parotidectomy was diminished in the EMG-monitored group ($p = 0.02$ and $p = 0.61$, respectively).

In a prospective study of 50 patients, Mamelle et al. [22] recently showed that supramaximal stimulation of the main trunk of the facial nerve at 2 mA at the beginning of dissection of the facial nerve and at the end of parotidectomy appears to be an effective parameter to predict immediate postoperative injury. The postdissection to predissection ratios of the maximal response amplitude, but not the stimulation thresholds, were significantly lower 2 days after surgery in the patient group with facial dysfunction compared to the group with a normal facial outcome (p < 0.02).

Liu et al. [28] analyzed 58 recurrent pleomorphic adenoma parotid surgeries performed between 2004 and 2012 to determine the benefit of intraoperative facial nerve monitoring during revision parotid surgery to postoperative facial function. There were no significant differences between the two groups in terms of the incidence of immediate or permanent facial paralysis after revision parotidectomy (p = 0.95 and p = 0.36, respectively). The average duration of surgery was longer and the severity of postoperative facial nerve palsy after total parotidectomy or wide resection was significantly worse without monitoring (p < 0.01 and p = 0.01, respectively). Moreover, the average recovery time for temporary facial nerve paralysis was significantly shorter in the monitored group compared to the unmonitored group, independent of the surgical technique (p < 0.01).

Pitfalls and Limitations

The surgeon or the neurophysiologist performing the monitoring must differentiate true EMG events from artifacts, such as those that occur from contact between surgical instruments in the operative field [29]. EMG waveform characteristics, the EMG amplitude, and the surgical context of the event aid in this differentiation. A false-positive event should be excluded because this may give the surgeon a false sense of insecurity. Too high a level of stimulation could result in a false-positive error, with inadvertent current spreading from the probe tip and through soft tissue around the site of injury, rather than through the injured nerve segment. To avoid this error, either a bipolar stimulator or only low-to-moderate current intensities with a fine-tipped monopolar stimulator should be used [30].

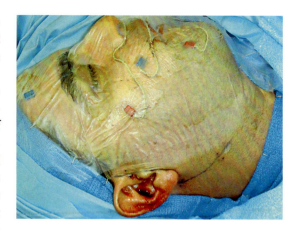

Fig. 4. Placement of the needle electrodes in the face for facial nerve monitoring. In this example, the recording needle electrodes are placed in the frontal muscle, orbicularis oculi muscle, orbicularis oris muscle, and mentalis muscle, corresponding to figure 1. It is important to note that the electrodes do not record from all facial muscles; rather, only the selected muscles and their corresponding facial nerve branches are recorded.

The absence of an electrically evoked response does not exclude the possibility that the stimulated tissue is the facial nerve [23]. In this situation, proper system operating function must be confirmed. A false-negative event may give the surgeon a false sense of security. The surgeon must exercise proper judgment in event interpretation, and anatomic information should always trump physiological information from the monitoring system. Electrosurgical units and other electrical equipment can create an electrical artifact that interferes with the detection of facial muscle responses. Thus, the use of electrosurgical dissection can create a gap in continuous facial nerve monitoring, and this limitation should be considered.

Appropriate electrode placement is very important. In parotid surgery, 2 or 4 channels are usually used for monitoring. For 2-channel monitoring, the needle electrodes are most frequently inserted into the superior portion of the orbicularis oris and orbicularis oculi muscles (fig. 4).

The surgeon must be aware that the monitoring system will only respond to evoked or mechanically induced responses of the muscles in which the recording electrodes are placed. The above-mentioned recording sites are sufficient to monitor the main trunk of the facial nerve but may lead to false-negative results if peripheral facial nerve branches that innervate other facial muscles are stimulated or injured. For this reason, some surgeons use 4-channel facial nerve monitoring in order to increase the monitoring field area.

Nerve Monitoring in Litigations Involving Parotid Surgery

A failure to use intraoperative facial nerve monitoring does not appear to be a relevant argument for the claim of medical malpractice in cases of salivary gland surgery with postoperative facial palsy. Overall, lawsuits related to parotid surgery are seldom [31]. Recently, Hong et al. analyzed United States civil trials involving medical malpractice in salivary gland surgery from 1985 to 2011. In no cases did a lack of facial nerve monitoring during parotid surgery contribute to the reason for the lawsuit [32]. The widely used treatise by Goldsmith on medical malpractice [33] stresses the controversial role of facial nerve monitoring in preventing facial nerve injuries and even suggests that the facial nerve monitor may only increase the cost and operative time [32]. Further experience with this methodology may change these perspectives.

Conclusion

Recent prospective trials have demonstrated that electrophysiological facial nerve monitoring can reduce the risk of early postoperative facial nerve dysfunction in primary parotid surgery, but not in revision surgery. The effect is more pronounced in total than in superficial parotidectomy. Monitoring is associated with a shorter surgical time in primary superficial parotidectomy compared to total parotidectomy. This seems to also be true for revision surgery. There is a lack of studies evaluating the long-term outcome using standardized assessments of facial nerve function. One study has shown that facial nerve stimulation at the end of parotidectomy compared to stimulation at the onset of surgery helps to prognosticate the facial nerve outcome. Additionally, it appears clear that facial nerve monitoring can help the surgeon to avoid facial nerve damage when the facial nerve is not formally dissected during parotid surgery, as for extracapsular tumor dissection or other minimally invasive parotid gland procedures. Further investigations are needed to demonstrate outcomes related to facial nerve monitoring during parotidectomy.

References

1 Nitzan D, Kronenberg J, Horowitz Z, et al: Quality of life following parotidectomy for malignant and benign disease. Plast Reconstr Surg 2004;114:1060–1067.
2 Ryzenman JM, Pensak ML, Tew JM Jr: Facial paralysis and surgical rehabilitation: a quality of life analysis in a cohort of 1,595 patients after acoustic neuroma surgery. Otol Neurotol 2005;26:516–521.
3 Guntinas-Lichius O, Gabriel B, Klussmann PJ: Risk of facial palsy and severe Frey's syndrome after conservative parotidectomy for benign disease: analysis of 610 operations. Acta Otolaryngol 2006;126:1104–1109.
4 Guntinas-Lichius O, Klussmann JP, Wittekindt C, et al: Parotidectomy for benign parotid disease at a university teaching hospital: outcome of 963 operations. Laryngoscope 2006;116:534–540.
5 Moeller K, Esser D, Boeger D, et al: Parotidectomy and submandibulectomy for benign diseases in Thuringia, Germany: a population-based study on epidemiology and outcome. Eur Arch Otorhinolaryngol 2013;270:1149–1155.
6 Sethi N, Tay PH, Scally A, et al: Stratifying the risk of facial nerve palsy after benign parotid surgery. J Laryngol Otol 2014;128:159–162.

7 Dillon FX: Electromyographic (EMG) neuromonitoring in otolaryngology-head and neck surgery. Anesthesiol Clin 2010;28:423–442.
8 Acioly MA, Liebsch M, de Aguiar PH, et al: Facial nerve monitoring during cerebellopontine angle and skull base tumor surgery: a systematic review from description to current success on function prediction. World Neurosurg 2013; 80:e271–e300.
9 Lalwani AK, Butt FY, Jackler RK, et al: Facial nerve outcome after acoustic neuroma surgery: a study from the era of cranial nerve monitoring. Otolaryngol Head Neck Surg 1994;111:561–570.
10 Schmitt WR, Daube JR, Carlson ML, et al: Use of supramaximal stimulation to predict facial nerve outcomes following vestibular schwannoma microsurgery: results from a decade of experience. J Neurosurg 2013;118:206–212.
11 Wilson L, Lin E, Lalwani A: Cost-effectiveness of intraoperative facial nerve monitoring in middle ear or mastoid surgery. Laryngoscope 2003;113:1736–1745.
12 Edwards BM, Kileny PR: Intraoperative neurophysiologic monitoring: indications and techniques for common procedures in otolaryngology-head and neck surgery. Otolaryngol Clin N Am 2005;38:631–642, viii.
13 Preuss SF, Guntinas-Lichius O: On the diagnosis and treatment of parotid gland tumors: results of a nationwide survey of ENT hospitals in Germany (in German). HNO 2006;54:868–874.
14 Hopkins C, Khemani S, Terry RM, et al: How we do it: nerve monitoring in ENT surgery: current UK practice. Clin Otolaryngol 2005;30:195–198.
15 Lowry TR, Gal TJ, Brennan JA: Patterns of use of facial nerve monitoring during parotid gland surgery. Otolaryngol Head Neck Surg 2005;133:313–318.

16 O'Regan B, Bharadwaj G, Elders A: Techniques for dissection of the facial nerve in benign parotid surgery: a cross specialty survey of oral and maxillofacial and ear nose and throat surgeons in the UK. Br J Oral Maxillofac Surg 2008;46: 564–566.
17 Sood AJ, Houlton JJ, Nguyen SA, et al: Facial nerve monitoring during parotidectomy: a systematic review and meta-analysis. Otolaryngol Head Neck Surg 2015;152:631–637.
18 Macdonald DB, Skinner S, Shils J, et al: Intraoperative motor evoked potential monitoring – a position statement by the American Society of Neurophysiological Monitoring. Clin Neurophysiol 2013;124:2291–2316.
19 Thiede O, Klusener T, Sielenkamper A, et al: Interference between muscle relaxation and facial nerve monitoring during parotidectomy. Acta Otolaryngol 2006;126:422–428.
20 Prell J, Rachinger J, Scheller C, et al: A real-time monitoring system for the facial nerve. Neurosurgery 2010;66: 1064–1073.
21 Prell J, Strauss C, Rachinger J, et al: Facial nerve palsy after vestibular schwannoma surgery: dynamic risk-stratification based on continuous EMG-monitoring. Clin Neurophysiol 2014;125:415–421.
22 Mamelle E, Bernat I, Pichon S, et al: Supramaximal stimulation during intraoperative facial nerve monitoring as a simple parameter to predict early functional outcome after parotidectomy. Acta Otolaryngol 2013;133:779–784.
23 Eisele DW, Wang SJ, Orloff LA: Electrophysiologic facial nerve monitoring during parotidectomy. Head Neck 2010;32: 399–405.
24 Schilling C, Gnanasegaran G, McGurk M: Three-dimensional imaging and navigated sentinel node biopsy for primary parotid malignancy: new application in parotid cancer management. Head Neck 2014;36:E91–E93.

25 Klintworth N, Zenk J, Koch M, et al: Postoperative complications after extracapsular dissection of benign parotid lesions with particular reference to facial nerve function. Laryngoscope 2010;120: 484–490.
26 Medina MV, Pollak N: Removal of an intra-parotid foreign body without parotidectomy and dissection of the facial nerve. Laryngoscope 2010;120(suppl 4): S132.
27 Grosheva M, Klussmann JP, Grimminger C, et al: Electromyographic facial nerve monitoring during parotidectomy for benign lesions does not improve the outcome of postoperative facial nerve function: a prospective two-center trial. Laryngoscope 2009;119:2299–2305.
28 Liu H, Wen W, Huang H, et al: Recurrent pleomorphic adenoma of the parotid gland: intraoperative facial nerve monitoring during parotidectomy. Otolaryngol Head Neck Surg 2014;151:87–91.
29 Hong RS, Kartush JM: Acoustic neuroma neurophysiologic correlates: facial and recurrent laryngeal nerves before, during, and after surgery. Otolaryngol Clin N Am 2012;45:291–306, vii–viii.
30 Kircher ML, Kartush JM: Pitfalls in intraoperative nerve monitoring during vestibular schwannoma surgery. Neurosurg Focus 2012;33:E5.
31 Hong SS, Yheulon CG, Wirtz ED, et al: Otolaryngology and medical malpractice: a review of the past decade, 2001–2011. Laryngoscope 2014;124:896–901.
32 Hong SS, Yheulon CG, Sniezek JC: Salivary gland surgery and medical malpractice. Otolaryngol Head Neck Surg 2013;148:589–594.
33 Goldsmith LS: Medical Malpractice: A Guide to Medical Issues. New York, Bender, 1986.

Orlando Guntinas-Lichius, MD
Department of Otorhinolaryngology, Jena University Hospital
Lessingstrasse 2
DE–07740 Jena (Germany)
E-Mail orlando.guntinas@med.uni-jena.de

Surgery for Benign Salivary Neoplasms

M. Boyd Gillespie[a] · Heinrich Iro[b]

[a]Department of Otolaryngology – Head and Neck Surgery, Medical University of South Carolina, Charleston, S.C., USA;
[b]Department of Otolaryngology – Head and Neck Surgery, University of Erlangen-Nuremberg, Erlangen, Germany

Abstract

Salivary neoplasms are relatively infrequent entities that account for only 4% of tumors of the head and neck. Although slow-growing lesions of the preauricular area and submandibular space are often confused with sebaceous cysts, lymph nodes, or lipomas by the non-otolaryngologist, otolaryngologists-head and neck surgeons recognize that all preauricular and submandibular masses should be considered a salivary neoplasm until proven otherwise. Surgery remains the treatment of choice for benign salivary gland neoplasms; however, techniques continue to evolve in order to preserve salivary function and reduce surgical morbidity. The goals of management of benign salivary neoplasms include accurate diagnosis of the lesion, complete surgical extirpation, and functional preservation of adjacent cranial nerves. Accurate diagnosis is aided by appropriate preoperative physical examination, imaging, and fine needle aspiration biopsy. Benign neoplasms typically present as slow-growing, painless, mobile masses without adverse features, such as tissue fixation, ulceration, a cranial nerve deficit, or regional lymphadenopathy. Preoperative imaging with ultrasonography, computed tomography, or magnetic resonance imaging reveals well-circumscribed lesions without an infiltrative growth pattern or associated adenopathy. Fine needle aspiration biopsy may favor a benign neoplasm, supporting the clinical presentation. Surgery for a benign or malignant salivary neoplasm is in essence a false dichotomy since the surgeon can never be completely confident of the diagnosis until the specimen is removed. The surgeon must recognize the significant overlap between benign and malignant salivary masses in terms of clinical presentation, imaging, and cytology, which requires the surgeon to remain vigilant and flexible at the time of surgery should tissue characteristics or frozen section analysis suggest a malignant process.

© 2016 S. Karger AG, Basel

Minor Salivary Gland

Presentation

Minor salivary gland neoplasms account for less than 15% of all salivary gland tumors. Most large case series have observed that 50% or more of these tumors are malignant [1–3]. Minor salivary neoplasms present as slowly enlarging, nonpainful submucosal masses that may be found incidentally or after they produce subtle sensations of

Table 1. Selection of appropriate surgical approach for benign parotid tumors based on tumor characteristics

	Tumor-related factors			
	Single lesion	Single lesion	Single lesion	Multiple lesions
Localization of tumor	Superficial	Superficial	Deep	Irrelevant
Mobility of tumor	Mobile	Fixed	Irrelevant	Irrelevant
Surgery of choice	Extracapsular dissection	Partial/superficial parotidectomy	Total parotidectomy	Total parotidectomy

pressure, fullness, or mild impairment in speech or swallowing. The hard and soft palate, buccal mucosa, and retromolar oral cavity are the most common sites. However, tumors may arise throughout the mucosa of the upper aerodigestive tract, including the paranasal sinuses, nasopharynx, oropharynx, larynx, trachea, and hypopharynx. The masses vary in consistency from firm to spongy on palpation. Adjacent cranial nerves should be assessed for any alteration in motor or sensory function.

Imaging
Although minor salivary tumors can be directly inspected transorally or with the aid of a fiberoptic scope, additional imaging is needed in order to determine the depth of the tumor and its relationship to surrounding structures. Computed tomography (CT) and magnetic resonance imaging (MRI) with intravenous contrast are the preferred methods of evaluation and may provide insight into the malignant potential of a given minor salivary gland mass. Fluorodeoxyglucose-positron emission tomography is rarely indicated in the primary evaluation of salivary tumors and cannot differentiate malignant potential due to significant uptake in both benign and malignant tumors [4]. Benign minor salivary gland tumors will display well-circumscribed borders that compress but do not infiltrate surrounding tissues. Bone remodeling and expansion may occur at the border, but bony irregularity and erosion raise concerns for malignancy. Although adenopathy is typically absent, reactive lymph nodes may be observed, especially following tumor biopsy.

Pathology
The most common benign neoplasms of the minor salivary glands are pleomorphic adenoma and monomorphic (e.g. basal cell) adenoma [5]. Most minor salivary masses are readily accessible for transoral fine needle aspiration (FNA) in order to confirm the diagnosis of salivary neoplasms. Due to significant overlap in cytologic features, it is often not possible to determine whether a given neoplasm is benign or malignant based on FNA alone [5]. Since this classification may impact the extent of surgical resection, an alternative option is core needle, punch, or wedge biopsy of the mass. These procedures can be performed more safely for minor salivary masses of the oral cavity and palate since major vessels and nerves are less proximal compared to masses of the parotid or submandibular gland. Due to their close proximity to the surface mucosa, incisional techniques can be performed with less risk of tumor spread or tracking; however, the area of the incision and any resulting scar must be subsequently removed with a sufficient margin at the time of definitive resection. Incisional biopsy techniques should target the border of the mass in order to examine the interface between the neoplasm and the surrounding tissues. Pathologic findings of surrounding tissue infiltration, perineural invasion, and/or lymphovascular invasion

provide evidence of a malignant process, whereas their absence suggests that the mass is likely benign.

Surgical Approach

Transoral or endoscopic approaches are preferred for minor salivary neoplasms of the upper aerodigestive tract that are judged to be benign based on appropriate preoperative examination, imaging, and pathology. Transoral and endoscopic approaches allow for complete removal of the mass, with a clear margin, while maximally preserving underlying tissues and nerves and avoiding the potential morbidity of transcervical or mandibulotomy approaches. Tumors of the oral cavity can be removed with the help of cheek and jaw retractors, whereas most tumors of the oropharynx are approachable with a tonsil gag. Tumors of the nasal cavity and nasopharynx can be removed using a transnasal endoscopic approach, whereas tumors of the tongue base, larynx, and hypopharynx require appropriate laryngoscopes. Although initially used for only squamous cell carcinoma, transoral robotic surgery is being increasingly used for other tumors of the upper aerodigestive tract [6]. This type of surgery provides excellent visualization and is capable of achieving negative margins in benign and early-stage (T1 and T2) malignant minor salivary gland neoplasms.

Magnified visualization via loupes, a telescope, or a microscope aids the dissection by making the border between tumor and normal tissue clearly visible. Once the tumor mass is sufficiently visualized, an adequate margin of dissection (5 mm for benign tumors) is marked around the border of the mass with an electrosurgical device. Resection the mucosa directly overlying the mass is preferred since attempts to preserve this layer may result in tumor spillage or a positive margin. Once the margin border is established, gentle retraction of the surface mucosa with sutures or clamps lifts the mass off the deep tissue plane, where a thin layer of normal tissue can be removed adjacent to the pseudocapsule in order to ensure a negative margin. Areas where the tumor is adherent require wider resection of adjacent tissue. Frozen section analysis can be performed on the excised mass if there is a concern about malignancy or on border areas where the tumor was adherent in order to confirm clear margins. Bleeding in the tumor bed can usually be controlled with oxymetazoline-soaked gauze and suction cauterization. Vessels larger than 3 mm may require control with surgical ties or clips. Defects less than 3 cm with an intact muscular base heal sufficiently with secondary intention. Split-thickness skin grafts or acellular dermal grafts may promote healing for oral cavity defects larger than 3 cm, especially in the buccal space or labial mucosa, where tissue retraction may occur (fig. 1). Full-thickness grafts or rotational flaps (e.g. uvula flap, palatal island flap) work best for the soft palate in order to preserve function and prevent retraction. Patients can be started on a soft diet immediately, with additional oral care using antiseptic oral rinses. Intraoral bolsters placed on grafts are removed within 1 week. Patients can proceed to their regular diet once the tissues are fully healed at 2–3 weeks postoperatively.

Sublingual Salivary Gland

Presentation and Evaluation

The evaluation and management of sublingual salivary gland masses are similar to those of minor salivary neoplasms. Sublingual gland neoplasms are rare entities (<5% of all salivary neoplasms) that are more likely to be malignant than any other salivary tumor, with rates of malignancy approaching 80% [7]. Sublingual neoplasms typically present as painless, mobile swellings in the floor of mouth. Fixation to the mandible or the floor-of-mouth musculature or changes in tongue sensation or mobility should be considered signs of malignancy until proven otherwise.

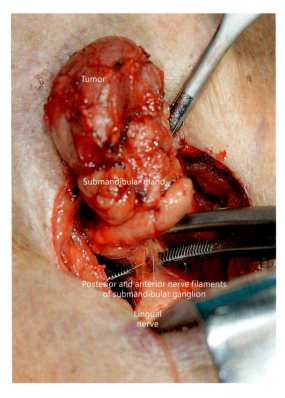

Fig. 1. Intraoperative view of submandibular gland excision, with exposure of the lingual nerve and the branches of the submandibular ganglion.

CT or MRI with contrast provides information with regard to the size of the mass and its relationship to the jaw, floor-of-mouth musculature, and submandibular gland and allows for an assessment of regional lymph nodes. Ultrasonography is challenging due to acoustic shadowing from the mandible. These neoplasms are readily accessible for FNA or core needle biopsy.

Surgical Approach

If preoperative evaluation by examination, imaging, and histopathology suggests a benign process, the surgery should focus on complete tumor removal, with preservation of adjacent floor-of-mouth structures, such as Wharton's duct and the lingual and hypoglossal nerves. The surgery can be performed transorally in almost all circumstances. Nasal intubation facilitates exposure to the floor of mouth and allows full mobility of the tongue, which can be easily retracted with a 2.0 silk stitch. A cheek and jaw retractor provides wide exposure to the floor of mouth. Magnification with loupes or an operating microscope allows for better visualization of the tissue planes. Cannulation of the ipsilateral Wharton's duct with lacrimal probes or salivary dilators prior to incision helps identification of this landmark during surgical dissection. The initial incision is made with a scalpel or electrocautery unit along the mucosal ridge just medial to Wharton's duct, where the ventral tongue merges with the floor of mouth. The incision should be 3–4 cm in length, from just behind the salivary ostium to the posterior floor of mouth, in order to provide full visualization of the sublingual gland, which is typically 3–4 cm in length and 1–2 cm in width. Gentle retraction on the lateral border of the incision with 4.0 Vicryl sutures lifts the sublingual gland and allows visualization of Wharton's duct, which runs along its medial surface. Blunt dissection with a Kitner sponge or mosquito forceps gently separates the sublingual gland from Wharton's duct. In the posterior floor of mouth, the large white fibers of the lingual nerve will be visualized as they cross underneath Wharton's duct from lateral to medial. Once the medial border of the gland is in view, an area of the gland not containing the mass can be grasped with an Allis or Babcock clamp. The mucosa overlying the lateral border of the gland can be dissected off the gland with dissection scissors or resected with the tumor if adherent. The gland is bluntly swept off the floor-of-mouth muscle, taking care to adequately cauterize penetrating veins and small arteries. External pressure on the submental and submandibular triangles often helps to lift the gland into the wound, thereby improving visualization. All sublingual tissue should be excised in order to avoid postoperative ranula. Once the gland is removed, the floor of mouth is irrigated, and the overlying mucosa is closed with interrupted Vic-

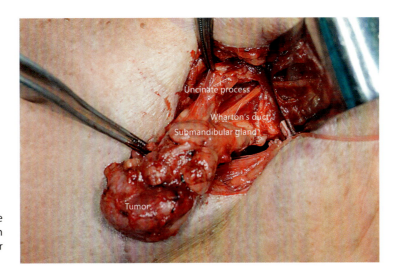

Fig. 2. Intraoperative view of the submandibular gland excision, with exposure of Wharton's duct after separation of the lingual gland.

ryl or chromic sutures. If there is insufficient mucosa for closure, a split-thickness skin graft or acellular dermal graft can be used if there is concern that the defect is too large to heal by secondary intention. Patients can be started on a soft diet immediately, with additional oral care using antiseptic oral rinses. Intraoral bolsters placed on grafts are removed within 1 week. Patients can proceed to their regular diet once the tissues are fully healed at 2–3 weeks postoperatively.

Submandibular Gland

Presentation
It is estimated that 10% of all salivary gland tumors, of which 40–50% are malignant, are located in the submandibular gland. The typical presentation of a benign tumor is a painless, slowly growing mass in the submental region, whereas fixation on the mandible and/or a paresis of the marginal branch of the facial nerve raises concern about malignancy.

Imaging
Diagnostic ultrasound is the imaging modality of choice. It can distinguish between a lump inside the submandibular gland and an impression of the gland by an external mass or lymph node. If there is a need to clarify if the lower jaw is infiltrated or if the tumor grows behind the mandible and toward the floor of mouth, CT should be preferred over MRI, especially if a malignant tumor is suspected. However, certain differentiation between benign and malignant tumors by means of diagnostic imaging is often not possible. As tumors of the submandibular gland are commonly located near the surface of the gland, FNA biopsy or core needle biopsy may be performed, recognizing the potential diagnostic challenges of salivary cytology.

Surgical Approach
The submandibular gland is usually extirpated through a horizontal skin incision parallel to the mandible. The glandular parenchyma (including the uncinated process) is completely removed, along with parts of the proximal Wharton's duct. Partial resections are not performed because of the risk of salivary fistula (fig. 1, 2).

In benign tumors, injuries to the marginal mandibular branch of the facial nerve, the lingual nerve, and the hypoglossal nerve should be avoided, as opposed to malignant tumors, where the

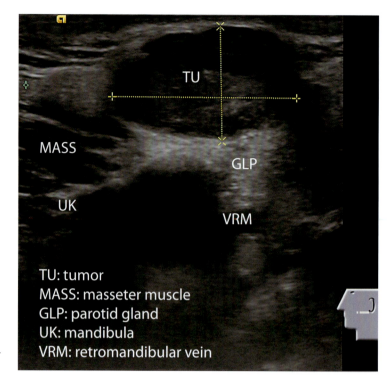

Fig. 3. Preoperative ultrasound image of a parotid gland tumor.

gland with infiltrated tissue, including involved nerves, should be removed. In those cases, the basic surgical procedure is a dissection of level I of the neck.

The complications after submandibular sialoadenectomy in benign tumors described in the literature include hemorrhage (0–14%), fistula (0–4%), postoperative skin infection (0–14%), transient facial palsy (10%), permanent facial palsy (3%), and permanent palsy of the lingual (2%) and hypoglossal (1%) nerves [8].

Parotid Gland

Presentation
Approximately 70% of all salivary tumors occur in the parotid gland, where 80% are in the superficial lobe, and 80% are benign [9]. These neoplasms typically present as painless, mobile swellings in the preauricular or infraauricular gland. Rapid growth; pain; facial weakness or twitching; facial numbness; skin fixation; reduced mobility; or prior skin cancers of the adjacent cheek, scalp, or temple should raise suspicion of a cancerous process. The extent of parotid gland removal depends on the clinical characteristics of the tumor (table 1).

Imaging
Ultrasonography is typically adequate for most masses of the superficial parotid. Benign lesions are usually well circumscribed, hypoechoic to surrounding tissues, and hypovascular, although these features are not specific to benign neoplasms (fig. 3). Measurement of the compressible features of the mass with sonoelastography indicates that malignant tumors are stiffer than benign tumors on average; however, this technique is not accurate enough to reliably differentiate be-

tween benign and malignant tumors [10]. Additional imaging with MRI or CT is reserved for situations where there is concern about malignancy, irregular or invasive features on ultrasound, or deep lobe or neck involvement. MRI is generally preferred over CT due to superior resolution of tissue planes within the parotid [11].

Pathology
The parotid gland has the most varied histopathology of all salivary glands, with pleomorphic adenoma, Warthin tumor, monomorphic adenoma, and oncocytoma being the most common subtypes. Most parotid masses are accessible to FNA biopsy with ultrasound guidance, which provides improved accuracy in sampling the target mass. Similar to other salivary tumors, there is a significant overlap between benign and malignant cytology features, so pathology can only be properly interpreted in light of the clinical presentation, radiographic features, and operative findings.

Surgical Approach: Superficial Parotidectomy
Prior to the mid-20th century, enucleation of parotid tumors was a common procedure, in which the capsule of a parotid tumor was opened and the contents of the tumor scooped out [12]. Enucleation was associated with recurrence rates of 20–45% since the procedure would leave behind the tumor capsule and increase the likelihood of tumor spillage into surrounding tissues. This high recurrence rate led to the surgical technique of superficial parotidectomy (SP), in which the outer part of the gland is removed by dissecting it free from the main trunk of the facial nerve and its divisions. SP reduced the rate of recurrence of pleomorphic adenoma and increased the time to recurrence from an average of 10 months to 7 years. Further support for SP includes the 10–15% probability that a given parotid mass is malignant, even when the clinical presentation, imaging, and cytopathology predict that it is benign. SP, however, results in significant dissection of the facial nerve along with loss of a large amount of glandular tissue and therefore has a higher rate of surgical complications and side effects, including permanent and temporary facial paralysis, Frey syndrome, and esthetic compromise due to loss of tissue volume. Use of facial nerve monitoring at the time of parotidectomy significantly decreases the rate of temporary nerve paresis but does not significantly decrease the long-term rate of permanent paralysis [13].

Partial Parotidectomy
Partial parotidectomy (PP) is a variation of SP that involves dissection of only the branches of the facial nerve in proximity to the tumor in order to remove the tumor along with a cuff of normal tissue. The procedure avoids direct exposure of the tumor capsule, if feasible, and therefore has the same low rates of tumor recurrence as SP, but with lower rates of facial nerve paresis and less loss of tissue volume. Whereas other techniques, such as extracapsular dissection (ED), require extensive judgement to determine whether a patient is a candidate, PP is applicable to most neoplasms of the superficial parotid, while providing adequate surgical excision in the event that a given mass is found to be malignant on final pathology. In addition, PP is easier to teach to surgical trainees since the only aspect of the surgery that varies from case to case is the facial nerve branches that require dissection.

Extracapsular Dissection

The surgical preparations do not differ between ED and conventional parotidectomy. Injection of a solution of 1:200,000 epinephrine in a local anesthetic is recommended prior to draping the patient because it provides an excellent bloodless field of dissection. The skin incision (according to Blair) is conducted around the ear lobe, starting at the tragus and ending in a skin fold in the neck;

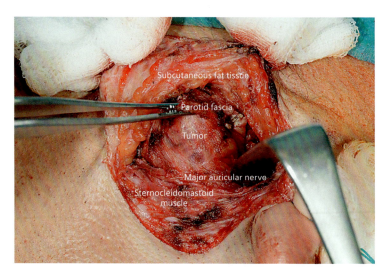

Fig. 4. Intraoperative view of parotid tumor extracapsular dissection, with the tumor exposed.

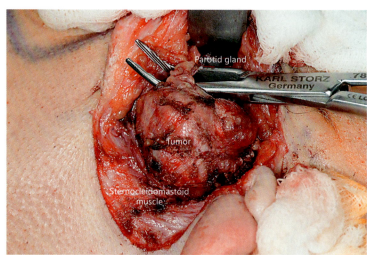

Fig. 5. Intraoperative view of parotid tumor extracapsular dissection, with extracapsular tumor preparation with curved forceps.

in contrast to conventional parotidectomy, the length of the incision and the resulting flap size in ED may be adapted to the size and location of the tumor. After dissection of the subcutaneous tissue, the sternocleidomastoid muscle, and the greater auricular nerve, the skin flap is raised, and the 'shining' capsule of the parotid gland is thus exposed (fig. 4). Before the capsule is opened, the tumor is once again palpated. If the exact position of the tumor cannot be determined, an ultrasound scan can be performed intraoperatively. The capsule of the parotid gland is now incised, and the dissection extended toward the tumor; however, the tumor capsule itself is never opened. Blunt dissection is now extended through the healthy glandular tissue around the tumor so as to gradually separate it, with care being taken at all times to dissect away from the tumor (fig. 5). With this technique, a small rim of healthy glandular tissue is left on the tumor, without damaging the facial nerve. Direct retraction of the tumor by instruments should be avoided in order to reduce the risk of rupture of the capsule; however, the covering parenchyma can be grasped gently.

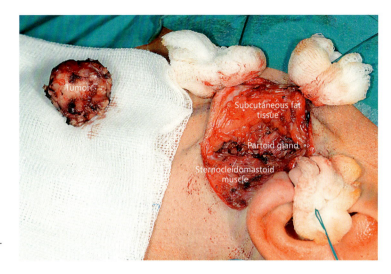

Fig. 6. View of the parotid gland excision site after tumor removal.

After the tumor has been removed, the defect within the parenchyma is checked for bleeding (fig. 6). The remnants of the edges of the parotid fascia are sutured back together, or in the case of a defect that cannot be closed primarily, it may be sutured to the anterior border of the sternocleidomastoid muscle. A rubber drain is inserted, and subcutaneous and skin sutures are applied. Finally, a pressure dressing may be applied.

Neuromonitoring and neurostimulation, consisting of an electrical bipolar stimulating probe and two electrodes for conducting the action potentials of the orbicularis oculi and orbicularis oris muscles, are mandatory. In ED, it is especially important to identify and protect visible branches of the facial nerve. The main trunk of the facial nerve, in contrast, is never exposed during ED. In cases where this is performed, the procedure changes to at least a PP.

Similar to permanent facial nerve paresis (0–5%), Frey syndrome (gustatory sweating) also occurs far less (0–5%) commonly after ED than after other techniques of parotidectomy.

ED should be performed by surgeons who are well experienced in all techniques of parotid gland surgery. It may be necessary to switch from ED to other surgical modalities due to the intraoperative situation.

References

1. Wang XD, Meng LJ, Hou TT, et al: Frequency and distribution pattern of minor salivary tumors in a northeastern Chinese population: a retrospective study of 485 patients. J Oral Maxillofac Surg 2015;73:81–91.
2. Ramesh M, Krishnan R, Paul G: Intraoral minor salivary gland tumours: a retrospective study from a dental and maxillofacial surgery centre in Salem, Tamil Nadu. J Maxillofac Oral Surg 2014;13:104–108.
3. Dalgic A, Karakoc O, Aydin U, et al: Minor salivary gland neoplasms. J Craniofac Surg 2014;25:e289–e291.
4. Toriihara A, Nakamura S, Kubota K, et al: Can dual-time-point 18F-FDG PET/CT differentiate malignant salivary gland tumors from benign tumors? AJR Am J Roentgenol 2013;201:639–644.
5. Turk AT, Wenig BW: Pitfalls in the biopsy diagnosis of intraoral minor salivary gland neoplasms: diagnostic considerations and recommended approach. Adv Anat Pathol 2014;21:1–11.
6. Villanueva NL, Almeida JR, Sikora AG, et al: Transoral robotic surgery for the management of oropharyngeal minor salivary gland tumors. Head Neck 2014;26:28–33.

7 Sun G, Yang X, Tang E, et al: The treatment of sublingual gland tumours. Int J Oral Maxillofac Surg 2010;39:863–868.
8 McGurk M, Makdissi J, Brown JE: Intraoral removal of stones from the hilum of the submandibular gland: report of technique and morbidity. Int J Oral Maxillofac Surg 2004;33:683–686.
9 Juengsomjit R, Lapthanasupkul P, Poomsawat S, et al: A clinicopathologic study of 1,047 cases of salivary gland tumors in Thailand. Quintessence Int 2015;46:707–716.
10 Westerland O, Howlett D: Sonoelastography techniques in the evaluation and diagnosis of parotid neoplasms. Eur Radiol 2012;22:966–969.
11 Shah GV: MR imaging of salivary glands. Neuroimaging Clin N Am 2004;14:777–808.
12 Albergotti WG, Nguyen SA, Zenk J, et al: Extracapsular dissection for benign parotid tumors: a meta-analysis. Laryngoscope 2012;122:1954–1960.
13 Sood AJ, Houlton JJ, Nguyen SA, et al: Facial nerve monitoring during parotidectomy: a systematic review and meta-analysis. Otolaryngol Head Neck Surg 2015;152:631–637.

M. Boyd Gillespie, MD, MSc, Professor
Department of Otolaryngology – Head and Neck Surgery
Medical University of South Carolina
135 Rutledge Ave. MSC 550
Charleston, SC 29425-5500 (USA)
E-Mail gillesmb@musc.edu

Recurrent Benign Salivary Gland Neoplasms

Robert Lee Witt[a] · Piero Nicolai[b]

[a]Department of Otolaryngology – Head and Neck Surgery, Christiana Care/Thomas Jefferson University, Newark, Del., USA;
[b]Department of Otorhinolaryngology – Head and Neck Surgery, University of Brescia, Brescia, Italy

Abstract

The most important causes of recurrence of benign pleomorphic adenoma are enucleation with intraoperative spillage and incomplete tumor excision in association with characteristic histologic findings for the lesion (incomplete pseudocapsule and the presence of pseudopodia). Most recurrent pleomorphic adenomas (RPAs) are multinodular. MRI is the imaging method of choice for their assessment. Nerve integrity monitoring may reduce morbidity of RPA surgery. Although treatment of RPA must be individualized, total parotidectomy is generally recommended given the multicentricity of the lesions. However, surgery alone may be inadequate for controlling RPA over the long term. There is growing evidence from retrospective series that postoperative radiotherapy results in significantly better local control. A high percentage of RPAs are incurable. All patients should therefore be informed about the possibility of needing multiple treatment procedures, with possible impairment of facial nerve function, and radiation therapy for RPA. Reappearance of Warthin tumor is a metachronous occurrence of a new focus or residual incomplete excision of all primary multicentric foci of Warthin tumor. Selected cases can be observed. Conservative surgical management can include partial superficial parotidectomy or extracapsular dissection. Not uncommonly, other major and minor salivary gland neoplasms, including myoepithelioma, basal cell adenoma, oncocytoma, canalicular adenoma, cystadenoma, and ductal papilloma, follow an indolent course after surgical resection, with rare cases of recurrence.

© 2016 S. Karger AG, Basel

Recurrent Pleomorphic Adenoma

Precise assessment of the recurrence rate of pleomorphic adenoma (PA) is difficult because of the small number of patients in most series and the variability in follow-up times. From 33 to 98% of recurrent pleomorphic adenomas (RPAs) are multifocal (fig. 1) [1, 2]. Many nodules are smaller than 1 mm and can be located in surrounding fat tissue. This explains the high incidence of second recurrence.

RPA tends to occur at a mean of 7–10 years after initial surgery. The time interval between intervention and recurrence is significantly shorter for enucleation [3] compared with more extensive procedures. The predisposing factors are variable, and the management of RPA is thus individualized.

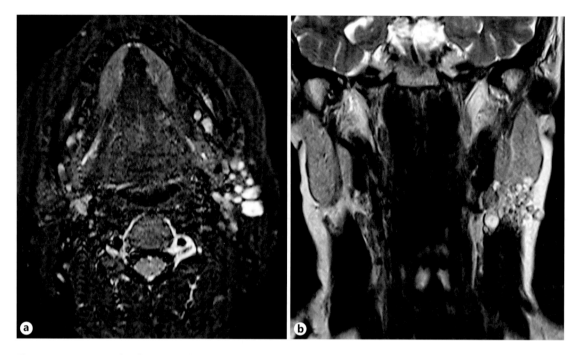

Fig. 1. Fat-suppressed turbo spin-echo T2-weighted sequence on axial plane (**a**) and turbo spin-echo T2-weighted sequence on coronal plane (**b**). Dissemination of small T2 hyperintense RPA nodules in the lower part of the superficial lobe of the left parotid gland in a 41-year-old woman who underwent enucleation. On the coronal plane, nodules in the subcutaneous fat tissue reaching the skin are also visible.

Etiology

The most important causes of recurrence are intraoperative rupture of the lesion with seeding of the surgical field, which increases the rate of recurrence by 5% compared to parotidectomy without tumor spill [4], and the use of enucleation, which is associated with a recurrent rate of as high as 45%, in contrast with the low rates of recurrence (1–4%) reported after superficial parotidectomy [4]. This remarkable difference is well explained by the focal absence of a capsule in PA. Therefore, dissection of the lesion from the adjacent salivary tissue can potentially detach small lobulations or pseudopodia of tumor cells outside of the presumed capsule and predispose the patient to tumor re-growth, which should be more appropriately indicated as a 'residual' rather than a 'recurrent' lesion.

Few studies have attempted to identify biologic predictors of recurrence. Brieger et al. [5] have found that cadherin 11, an adhesion protein, and fascin, a bundling protein with possible roles in cell motility and detachment of tumor cells, are both over-expressed in tumors that subsequently recur. More recently, de Souza et al. [6] performed immunohistochemical analyses to investigate the expression of cell cycle proteins in PA, RPA, and carcinoma ex-pleomorphic adenoma (CXPA). In contrast with PA, RPA and CXPA demonstrated strong staining for p16, cyclin D1, CDK4, E2F, and retinoblastoma protein. A similar distribution in PA, in contrast with RPA and CXPA, was observed for polypeptides belonging to the family of platelet-derived growth factors and fibroblast growth factors, thus suggesting their potential roles in recurrence and malignant transformation of PA [7].

RPA has been reported to occur more frequently in tumors that are hypocellular and chondromyxoid in nature with a higher rate of incomplete encapsulation [8]. The mean age at initial presentation of PA among patients who later develop recurrence is significantly lower than the mean age of those who remain free from disease at long-term follow-up. Female gender is also a reported risk factor [9].

Enucleation of upper cervical 'lymph nodes' with a final pathological diagnosis of PA can lead to RPA, as can 'incision and drainage' of a suspected peri-tonsillar abscess that is parapharyngeal PA. Open biopsy or core needle biopsy can also lead to RPA.

Treatment Planning, Surgical Strategy, and Outcome

Management of RPA is a challenge for head and neck surgeons. The imaging method of choice is MRI, which is more accurate than ultrasound or CT in determining the amount of residual parotid gland, the multinodular nature of disease, the number of nodules, and potential deep-lobe or parapharyngeal extension.

Since the tendency of the lesion is to grow, although at an unpredictable rate, and the risk of malignant transformation increases with time, surgery is generally advised. Irrespective of the type of resection elected, removal of the skin scar from a previous surgery and the use of loop or surgical microscopic magnification to identify small nodules below the resolution of MRI and to assist the surgeon in tracing the facial nerve and its main branches are strongly recommended. A wait-and-see policy may be followed for the elderly or for patients with severe co-morbidities.

Owing to the fact that no standardized guidelines for RPA surgery are available and no prospective studies have compared the extent of surgery and outcome, treatment must be individualized based on the age of the patient, presentation (uninodular vs. multinodular) and location (superficial vs. deep) of the lesion, number and types of previous surgeries, and presence of residual salivary parenchyma. As a general recommendation, RPA surgery should not be attempted by a novice or occasional salivary gland surgeon.

Total parotidectomy has been favored for either presumed uninodular or multinodular recurrence after superficial parotidectomy [10] because of the frequent presence of multinodularity. Revision surgery by total parotidectomy will result in a high rate of transient and permanent facial nerve dysfunction. Total parotidectomy may reduce but does not prevent the leaving of microscopic residuals. The re-recurrence rate after 15 years has been reported to be 75% for extended parotidectomy that is not followed by radiotherapy (RT) [9].

A single superficial recurrent lesion detected by preoperative MRI (fig. 2) and intraoperative evaluation occurring after enucleation or limited superficial parotidectomy does not necessarily require total parotidectomy. It is recommended to resect scars from any previous incisions.

Localized resection of a tumor can be employed after multiple recurrences or after previous total parotidectomy when it is the only option to preserve the facial nerve. A higher failure rate is expected to result from re-excisional biopsy than from formal parotidectomy for RPA [11].

Radical and extended parotidectomies have been considered for patients with infiltration of branches or the main trunk of the facial nerve. The need to sacrifice a branch of the facial nerve during RPA surgery has been reported in 14–30% of patients [2, 12]. Facial nerve resection may be necessary for patients with a history of multiple recurrences or failed RT. Even total parotidectomy with facial nerve sacrifice does not prevent further recurrence.

When multinodular recurrences involve subcutaneous tissues of the upper neck, compartmental resection superficial to the platysma and

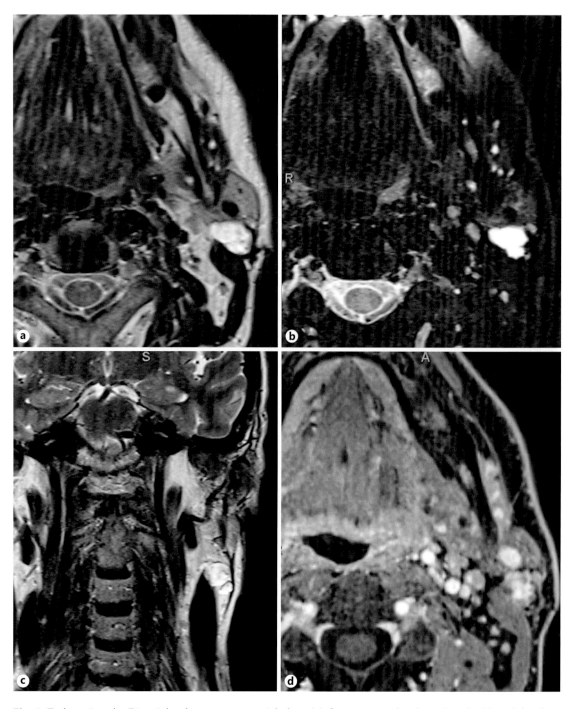

Fig. 2. Turbo spin-echo T2-weighted sequence on axial plane (**a**), fat-suppressed turbo spin-echo T2-weighted sequence on axial plane (**b**), turbo spin-echo T2-weighted sequence on coronal plane (**c**), and fat-suppressed T1 after contrast injection on axial plane (**d**). A uninodular RPA is identified in contact with the residual parotid tissue in a 47-year-old man who underwent enucleation 7 years earlier.

including the fat tissue potentially harboring micro-lesions can minimize the chance of recurrence. Parapharyngeal RPA is associated with a lower chance of curative treatment [12].

The surgical goal for RPA management is maintenance of a balance between the conflicting goals of complete surgical resection of what is often multicentric RPA and quality of life with facial nerve preservation. The ultimate control rates for first revision surgery alone have been reported to range from 36 to 98% [2, 9, 13]. Control rates vary depending on the length of follow-up and can be difficult to accurately measure as patients can be lost to follow-up.

The recurrence rate following a second recurrence has been reported to be 45% [1] after 10 years of follow-up. Recurrence rates are higher for previously recurrent tumors than for those that develop after the first operation. Redaelli de Zinis et al. [2] have observed that predictors of recurrence are local excision vs. formal parotidectomy and the presence of multinodular disease.

Surgical Morbidity: Prevalence and Prevention

The incidence of postoperative facial paralysis after primary parotid surgery has been reported to be 9.1–64.0% for temporary facial paralysis and 0–3.9% for permanent facial paralysis [4]. The rates of temporary and permanent facial nerve dysfunction are considerably higher after RPA parotid surgery and have been reported to be 90–100% and 11.3–40.0%, respectively [1, 9]. The rate of facial nerve dysfunction increases with each revision procedure. Previous dissection makes it difficult to distinguish the facial nerve from scar tissue. Retrograde facial nerve dissection starting from the more likely undisturbed temporal facial nerve branch has an important potential role in RPA surgery. Another concern following tumor recurrence relates to a facial nerve that is closely adherent to a tumor. Sacrifice of the facial nerve with microsurgical nerve grafting has been reported to occur in 14–30% of patients with RPA [2, 14].

In RPA surgery, unlike primary surgery, the surgeon does not always skeletonize nerve branches but instead attempts to preserve the fibrosis that surrounds the branches, which limits nerve exposure and mechanical manipulation. By avoiding devascularization and perineurium exposure, removal of RPA can be achieved after identification of the nerve branches via nerve monitoring without exposing the nerve. This strategy potentially reduces surgical trauma caused by mechanical separation.

Neuromonitoring has been reported to significantly reduce the operative time, decrease the incidence rate of permanent facial nerve paralysis, and shorten the recovery time of postoperative facial nerve function [12]. A reduction in the severity of facial nerve injury [15] also has been reported.

Radiotherapy

The indications for adjuvant RT in RPA are among the most debated issues in management of the lesion. Although a lack of randomized studies prevents the development of recommendations with a high level of evidence, the results of recent studies [16, 17] have suggested that RT is indicated when complete extirpation of the disease is not possible, when sacrifice of the facial nerve would otherwise have been necessary, in cases of multinodular recurrence, after multiple recurrences, or when further recurrence is likely to result in significant damage to the facial nerve. RT after surgery is not necessarily indicated for an isolated recurrence in a younger patient.

Given the potentially high rate of re-recurrence of RPA even with total parotidectomy and extended parotidectomy, the surgeon should discuss the option of adjuvant RT with the patient. The potential risks of a future redo surgery should be weighed against the small but real risk of radiation-induced malignancy. Gross residual disease is unlikely to be cured by RT alone [16].

Carcinoma Ex-Pleomorphic Adenoma

The de novo malignant transformation of RPA has been reported in 0–23% of cases [1, 9]. The risk of malignant transformation may increase with time and with the number of recurrences and may be preceded by PA with dysplastic change.

Metachronous Warthin Tumor

Etiology
The pathophysiology of Warthin tumor (WT) is markedly different from those of PA and other salivary gland neoplasms, both benign and malignant. WT is unique to humans and is associated with smoking in 90% of cases [18, 19]. WT is not a clonal neoplasm and is more accurately described as papillary cysto-lymphomatosis. Reappearance of WT is a metachronous occurrence of a new neoplastic focus or residual incomplete excision of all primary multicentric foci of WT. This can occur in 5–10% of cases [20].

Diagnosis
Inflammatory lymphoid stroma can result in a clinical presentation of lower parotid or upper neck area pain with primary or metachronous WT (MWT). WT presents in the tail of the parotid in 78–96% of patients [19, 21], where the greatest distribution of intra-parotid lymph nodes is present [22]. MRI can best define an MWT.

MWT can be diagnosed by cytology with a high degree of certainty if the diagnostic triad of oncocytes, lymphoid cells, and a proteinaceous background are present. In a considerable number of smears, the aspirate contains only two or a single component. Fine-needle aspiration (FNA) cytology of WT may be nondiagnostic if only the cyst contents are aspirated. Ultrasound-guided FNA, ideally with the cytopathologist present, will potentially reduce sampling error. If a lymphocyte-rich area is aspirated, an erroneous diagnosis of a lymphoproliferative lesion such as lymphoma may be considered, but this diagnosis can be excluded with flow cytometry of the FNA aspirate. If predominantly oncocytic epithelium is sampled, then it may be difficult to distinguish WT from oncocytoma. A frozen section diagnosis can be confirmed by histology in up to 100% of cases [19].

A parotid mass in a patient with prior WT, a smoking history, pain or cellulitis of the lower pole of the parotid/upper neck, and an FNA diagnosis of WT is highly likely to have a frozen section and final pathological diagnosis of MWT.

Management
For MWT, a 'wait-and-see' approach can be offered to select patients after a near certain clinical and FNA cytology diagnosis of WT is made. Surgical management of patients with likely MWT is an appropriate option in many cases given the high rates of pain and cellulitis [19]. A less invasive parotid surgical approach of partial superficial parotidectomy (PSP) or extra-capsular dissection (ECD) can be contemplated with confidence when supported by the clinical presentation, benign FNA cytology, and frozen section diagnosis of WT. Given that WT and MWT are generally located in the inferior pole, total parotidectomy or complete superficial parotidectomy is generally not warranted, even when attempting to encompass occult multicentric MWT. PSP and ECD are more focused and less invasive surgical procedures. Limited dissection of the buccal, orbital, and frontal nerve branches of the facial nerve can reduce morbidity. Nerve integrity monitoring is recommended for revision surgery for MWT. Equivalent outcomes have been reported for ECD and PSP for WT [19].

Other Recurrent Benign Salivary Neoplasms

Other relatively common major and minor salivary gland neoplasms, including myoepithelioma, basal cell adenoma, oncocytoma, canalicular adenoma, cystadenoma, and ductal papilloma, largely follow an indolent course after surgical resection with rare cases of recurrence. Exceptions are the rare parotid multifocal membranous subtype of basal cell adenoma, multifocal minor salivary gland oncocytoma, and sialadenoma papilliferum [23–25].

Recurrent Benign Salivary Neoplasms of the Submandibular Gland and Minor Salivary Glands

PA arises <10% of the time in the submandibular glands [26] and minor salivary glands [26], and evidence-based practice data are thus sparse. Empirically, the etiology, preoperative evaluation and management would follow the paradigm outlined for recurrent parotid PA. This would include resection of old scar tissue, as well as landmark identification, including the digastric muscle, mylohyoid muscle, marginal mandibular branch of the facial nerve, hypoglossal nerve, lingual nerve, jugular vein, carotid artery, and vagus nerve.

Prevention of recurrence for neoplasms of the submandibular triangle could potentially be enhanced using an initial procedure that involves regional dissection of the submandibular triangle, including the entire gland and upper cervical lymph nodes [27]. Other groups have proposed trans-cervical local excision of tumors with limited tumor-free margins and preservation of the remnant glandular tissues based on no recurrence over a medium period of 36 months [28] and even trans-oral approaches [29].

Treatment of recurrent PA of the submandibular gland should include wide resection of the submandibular triangle to encompass multifocal disease coupled with selective neck dissection, including levels Ib, IIa, and III [27].

Most PAs of minor salivary glands are located in the palate, followed by the buccal mucosa and lip. They are not as commonly multicentric. Patients with recurrent minor salivary gland PA should undergo imaging and biopsy to exclude multicentricity and malignant transformation [30]. Wide resection with a margin of normal surrounding tissue will reduce the risk of re-recurrence.

Conclusions

The most important cause of recurrence of PA is enucleation with rupture and incomplete tumor excision. RPA is often multicentric. Surgical treatment is individualized, with accumulating evidence from retrospective series indicating that postoperative RT results in significantly better local control. Reappearance of WT is a metachronous occurrence of a new focus or residual incomplete excision of all primary multicentric foci of WT. Selected cases can be followed by observation. Conservative surgical management can include PSP or ECD.

References

1 Zbären P, Tschumi I, Nuyens M, et al: Recurrent pleomorphic adenoma of the parotid gland. Am J Surg 2005;189:203–207.
2 Redaelli de Zinis LO, Piccioni M, Antonelli AR, et al: Management and prognostic factors of recurrent pleomorphic adenoma of the parotid gland: personal experience and review of the literature. Eur Arch Otorhinolaryngol 2008;265: 447–452.
3 Niparko JK, Beauchamp ML, Krause CJ, et al: Surgical treatment of recurrent pleomorphic adenoma of the parotid gland. Arch Otolaryngol Head Neck Surg 1986;112:1180–1184.
4 Witt RL: The significance of the margin in parotid surgery for pleomorphic adenoma. Laryngoscope 2002;112:2141–2154.
5 Brieger J, Duesterhoeft A, Brochhausen C, et al: Recurrence of pleomorphic adenoma of the parotid gland – predictive value of cadherin-11 and fascin. APMIS 2008;116:1050–1057.
6 de Souza AA, Altemani A, Passador-Santos F, et al: Dysregulation of the Rb pathway in recurrent pleomorphic adenoma of the salivary glands. Virchows Arch 2015;467:295–301.

7 Soares AB, Demasi AP, Tincani AJ, et al: The increased PDGF-A, PDGF-B and FGF-2 expression in recurrence of salivary gland pleomorphic adenoma. J Clin Pathol 2012;65:272–277.

8 Stennert E, Guntinas-Lichius O, Klussmann JP, et al: Histopathology pleomorphic adenoma in the parotid gland: a prospective unselected series of 100 cases. Laryngoscope 2001;111:2195–2200.

9 Wittekindt C, Streubel K, Arnold G, et al: Recurrent pleomorphic adenoma of the parotid gland: analysis of 108 consecutive patients. Head Neck 2007;29:822–828.

10 Stennert E, Wittekindt C, Klussmann JP, et al: Recurrent pleomorphic adenoma of the parotid gland: a prospective histopathological and immunohistochemical study. Laryngoscope 2004;114:158–163.

11 Niparko JK, Kileny PR, Kemink JL, et al: Neurophysiologic intraoperative monitoring: II. facial nerve function. Am J Otol 1989;10:55–61.

12 Makeieff M, Venail F, Cartier C, et al: Continuous facial nerve monitoring during pleomorphic adenoma recurrence surgery. Laryngoscope 2005;115:1310–1314.

13 Maran AG, Mackenzie IJ, Stanley RE: Recurrent pleomorphic adenoma of the parotid gland. Arch Otolarynogol 1984;110:167–171.

14 Conley J, Clairmont AA: Facial nerve in recurrent benign pleomorphic adenoma. Arch Otolaryngol 1979;105:247–251.

15 Liu H, Wen W, Huang H, et al: Recurrent pleomorphic adenoma of the parotid gland: intraoperative facial nerve monitoring during parotidectomy. Otolaryngol Head Neck Surg 2014;151:87–91.

16 Samson MJ, Metson R, Wang CC, et al: Preservation of the facial nerve in management of recurrent pleomorphic adenoma. Laryngoscope 1991;101:1060–1062.

17 Carew JF, Spiro RH, Singh B, et al: Treatment of recurrent pleomorphic adenomas of the parotid gland. Otolaryngol Head Neck Surg 1999;121:539–542.

18 Vories A, Ramirez S: Warthin's tumor and cigarette smoking. South Med J 1997;90:416–418.

19 Witt RL, Iacocca M, Gerges F: Contemporary diagnosis and management of Warthin's tumor. Del Med J 2015;87:13–16.

20 Klussmann JP, Wittekindt C, Preuss SF, et al: High risk for bilateral Warthin tumor in heavy smokers-review of 185 cases. Acta Otolaryngol 2006;126:1213–1217.

21 Garatea-Crelgo J, Gay-Escoda C, Bermejo B, et al: Morphological study of the parotid lymph nodes. J Craniomaxillofac Surg 1993;21:207–209.

22 Parwani A, Ali S: Diagnostic accuracy and pitfalls in fine-needle aspiration interpretation of Warthin's tumor. Cancer 2003;99:166–171.

23 Yu G, Ubmuller J, Donath K: Membranous basal cell adenoma of the salivary gland: a clinicopathologic study of 12 cases. Acta Otolaryngol 1998;118:588–559.

24 Martin H, Janda J, Behrbohm H: Locally invasive oncocytoma of the paranasal sinuses (in German). Zentralbl Allg Pathol 1990;136:703–706.

25 Brannon RB, Sciubba JJ, Guilani M: Ductal papillomas of salivary gland origin: a report of 19 cases and review of the literature. Oral Surg Oral Med Oral Pathol Oral Radiol Endod 2001;92:68–77.

26 Alves FA, Perez DE, Almeida OP, et al: Pleomorphic adenoma of the submandibular gland. Arch Otolarynol Head Neck Surg 2002;128:1400–1403.

27 Munir N, Bradley PJ: Diagnosis and management of neoplastic lesions of the submandibular triangle. Oral Oncol 2008;44:251–260.

28 Ro JL, Park CI: Gland-preserving surgery for pleomorphic adenoma in the submandibular gland. Br J Surg 2008;95:1252–1256.

29 Hong KH, Yang YS, Hong KH, et al: Intraoral approach for the treatment of submandibular salivary gland mixed tumors. Oral Oncol 2008;44:491–495.

30 Karatzanis AD, Drivas EI, Giannikaki ES, et al: Malignant myoepithelioma arising from recurrent pleomorphic adenoma of the soft palate. Auris Nasus Larynx 2005;32:435–437.

Robert Lee Witt, MD, FACS
Christiana Care/Thomas Jefferson University
4745 Ogletown-Stanton Rd, MAP #1, Suite 112
Newark, DE 19713 (USA)
E-Mail RobertLWitt@gmail.com

Prognostic Scoring for Malignant Salivary Gland Neoplasms

Vincent Vander Poorten[a] · Orlando Guntinas-Lichius[b]

[a] Otorhinolaryngology and Head and Neck Surgery, Section Head and Neck Oncology, Department of Oncology, University Hospitals Leuven, Leuven, Belgium; [b] Department of Otorhinolaryngology, Institute of Phoniatry/Pedaudiology, Jena University Hospital, Jena, Germany

Abstract

Estimating the prognosis of a patient with a rare disease like salivary gland carcinoma has always been a significant challenge for the clinician. Recent evolution in prognostic research has resulted in systems that summarize the effect of properly weighted, multivariate independent prognostic factors into a scientifically based prognostic estimate. Following external validation and user-friendly translation, these systems constitute an advance in clinical practice and have the ability to influence clinical decision making for the oncologist and the patient.

© 2016 S. Karger AG, Basel

Introduction

When confronted with a patient with a primary diagnosis of salivary gland carcinoma (SGC), it is difficult for the clinician to predict the patient's chances of cure and survival for 3 obvious reasons. First, no clinician really has extensive experience. The threshold for defining rare disease, i.e. <6 new cases per 100,000 patients per year (http://www.rarecare.eu), is still ten times the European incidence of 6–7 new SGC cases per 10^6 patients per year [1]. In total, 70% of SGCs arise in the parotid; 10–25%, in the minor glands; and the remainder, mainly in the submandibular glands [1]. Second, within this rare disease, 24 histological subtypes have been defined, and recently, newer entities have been described (e.g. mammary secretory analog carcinoma [2]). Even then, tumors of the same subtype show diverging clinical behavior, which is the third complicating factor [3]. Histopathological grading of SGC attempts to explain this variable biology within one histotype, but unfortunately, there remains a lack of uniformity in applying the grading criteria [4]. Collinearities between histological grade and the more easily assessable factors (stage, R1 resection) often makes 'grading' less essential in decision making than its coinciding factors [5]. Grading is routinely performed for the three most frequent histotypes: acinic cell carcinoma (AcCC) [6] mucoepidermoid carcinoma (MEC) [7] and adenoid cystic carcinoma (ACC) [8]. Treatment implications currently result only from recognizing non-high-grade MEC and acinic cell carcinoma, as low-stage, completely resected tumors may not require adjuvant radiation therapy [7, 9].

For any specific SGC patient, the aim is to predict the result of the anticipated treatment outcome compared to the 'overall outcome' of the entire group. Table 1 displays such an overall outcome in terms of disease-specific survival (DSS) for the 'parotid carcinoma group' [5]. These numbers are too general to be used for counseling any specific patient, with each patient having a unique set of prognostic factors that jointly imply a better or worse outcome than the whole group's prognosis. What follows is a review of some of the major research efforts to assess SGC prognosis. These efforts mainly aim at quantifying the contribution of the different factors to prognosis and summarizing each factor, with its respective weight, into a single number that reflects the individual SGC patient's prognosis.

General Issues: Prognostic Research Relates Oncological Outcomes to Specific Prognostic Factors

Oncological outcomes are related to time-event outcomes, which combine a time variable (the interval between diagnosis and the date of last follow-up) with an indicator, whether the outcome studied occurred at that last follow-up or not (censoring). The outcome can vary, e.g. 'death from any cause' when studying 'overall survival' or 'local recurrence, regional recurrence or recurrence at a distance or a combination' when studying the 'recurrence-free interval'. The latter outcome reflects the expected treatment result following initial therapy. When the focus is on 'DSS', death due to the tumor is the central event, and DSS reflects the best obtainable result for all possible treatments given. The event is called a 'failure' because it is what treatment aims to avoid. The specific prognostic factors linking to the above outcomes in SGC are patient-, tumor- and treatment-related clinical, pathological and increasingly molecular biological factors [5].

Prognostic Research Echelons in Salivary Gland Carcinoma, with Increasing Complexity Paralleling Increasing Clinical Usefulness

These echelons consist of (1) univariate and (2) multivariate identification of prognostic factors (see next section); (3) combining these factors into a summary index, score or nomogram, providing a user-friendly tool (see section Creation of a Summary Index, Score or Nomogram); and (4) external validation. The last proves applicability outside the research population that was used to construct the tool (see section The Last Step on the Prognostic Ladder: External Validation). As specified in the following paragraphs, depending on the research group, these efforts have focused on parotid carcinoma alone or on the entire group of SGCs at different anatomical sites.

Univariate and Multivariate Survival Analyses for Patients with Parotid Carcinoma

Many prognostic factor studies have used univariate Kaplan-Meier survival analysis, which has only relative practical utility in the clinical setting [11]. Cox proportional hazards multivariate survival analysis takes us one step closer to clinical usability [11] because the impact of one prognostic factor is adjusted for the effect of other factors [5]. These approaches identified (1) patient factors (age, gender, pain and comorbidity); (2) tumor factors (histological type, grade, stage, skin and soft tissue invasion, facial nerve involvement and perineural growth); and (3) treatment factors (resection margins and adjuvant radiotherapy) as important prognostic factors [5]. Sadly, the clinician seeking to apply this information when counselling a patient is left helpless with a table listing the prognosticators together with accompanying p values or hazard ratios (HRs) with confidence intervals (CIs), reflecting the prognosticators' respective importance. To be able to provide a prognostic estimate for a patient with SGC, the clini-

Table 1. DSS for parotid carcinoma

Research group	Publication year	Patients, n	5-year DSS, %	10-year DSS, %
Spiro [30]	1986	623	55	47
Spiro et al. [31]	1989	62	63	47
Kane et al. [32]	1991	194	69	68
Poulsen et al. [33]	1992	209	71	65
Leverstein et al. [34]	1998	65	75	67
Therkildsen et al. [35]	1998	251	76	72
Renehan et al. [36]	1999	103	78	65
Vander Poorten et al. [13]	1999	168	59	54
Harbo et al. [37]	2002	152	57	51
Godballe et al. [38]	2003	85	52	
Vander Poorten et al. [14]	2003	231	62	
Lima et al. [39]	2005	126	72	69
Mendenhall et al. [40]	2005	224		57
Vander Poorten et al. [20]	2009	237	69	58
Guntinas-Lichius et al. [10]	2015	295	82	82

Modified from and reprinted with permission from Vander Poorten et al. [5].

cian has to intuitively amalgamate these factors, which is confused by the facts that some factors are retained in one study but rejected in another and that different studies focus on differently composed study groups and investigate different outcomes. The conclusion arrived at in clinical practice is that most clinicians do not use the information resulting from these first two steps on the prognostic ladder when counseling patients with SGC [12].

One Step Further: Creation of a Summary Index, Score or Nomogram

How to Interpret p Values and Hazard Ratios
The suggestion that a smaller p value or a narrower CI accompanying a prognosticator means that this factor is prognostically more important than another with a larger p value is false. The magnitude of these p values depends heavily on the definition used, the number of the factors studied, and the patient number as well as on the sequence in which the variables are entered into the Cox model. If one of these aspects changes, the p value will change accordingly. The important mutual effect that specific variables have when going from one step to the next in a multivariate model while introducing new variables is usually not reported. In a multivariate analysis of the patient group of the Netherlands Cancer Institute, we insisted on including (1) the exact definition of the variables and their scale as well as (2) the exact sequence of entry of all variables. This is the only way that the reader can, at every step in stepwise model generation, appreciate the p values and their change after entering the next variable into the model [13]. These p value changes upon the addition of a new variable to an existing set of variables occur when the already included variables and the newly entered one refer to the same aspect of the risk of treatment failure. It is thus understandable that a given prognostic variable (e.g. clinical grade of malignancy) that is important in one study does not appear in the multivariate model in another study where the final model is composed of other strongly associated variables (e.g. in the case of the clinical grade

Table 2. Identified prognostic factors: coefficients and HRs in parotid cancer [13]

	Coefficient	HR (95% CI)	p value[a]
First model: variables in the pre-therapeutic analysis			
Pain on presentation	0.62	1.85 (1.028–3.33)	0.040
Age at diagnosis (linearly)	0.024	1.024 (1.007–1.042)	0.006
T classification (linearly)	0.44	1.55 (1.10–2.19)	0.012
N classification (linearly)	0.45	1.57 (1.28–1.94)	<0.001
Skin invasion	0.63	1.87 (0.97–3.61)	0.061
Facial nerve dysfunction	0.91	2.49 (1.31–4.72)	0.005
Second model: variables in the post-therapeutic analysis			
Age at diagnosis (linearly)	0.018	1.018 (1.001–1.034)	0.037
T classification (linearly)	0.39	1.47 (0.99–2.20)	0.055
N classification (linearly)	0.34	1.40 (1.11–1.76)	0.004
Skin invasion	0.70	2.00 (1.04–3.85)	0.037
Facial nerve dysfunction	0.56	1.75 (0.88–3.45)	0.109
Perineural growth	0.78	2.19 (1.05–4.59)	0.038
Positive surgical margins	0.65	1.91 (0.92–3.98)	0.083

[a] The p value is a way to quantify the exactness of the estimated coefficient and gives the chance to observe the displayed HR when the real value is 1 (or there is no increased associated risk of recurrence). Cancer©, copyright 1999, American Cancer Society.

of malignancy, this can be age, TNM classification, or positive resection margins). This simply means that the levels of the factor 'clinical grade of malignancy' divide the patient group into the same sets of prognostic groups as the TNM classification, the age groups, or the negative and positive resection margin groups do. In multivariable analysis, the aim is to define the most parsimonious set of prognosticators that best explains the observed outcome variability, rather than to define the one most important prognostic factor based on the smallest p value.

Creation of a Summary Index, Score or Nomogram

The abovementioned 'set of prognosticators' can be summarized in a prognostic index, score or a nomogram and externally validated in different populations. A practical, easy-to-use translation is a prerequisite for real use of the summary measure in routine clinical practice.

Development of a Prognostic Index for Patients with Parotid Carcinoma [13]

A total of 151 consecutive patients of the Netherlands Cancer Institute who underwent primary curative treatment for parotid carcinoma between 1973 and 1994 (median follow-up 94 months) were subjected to a multivariate analysis of clinical and pathological factors to define the set of prognosticators that best explained the observed variability in the 'recurrence-free interval'. The factors studied were highly correlated (e.g. perineural growth associated with a higher percentage of positive margin resections), so a stepwise Cox model was the best approach to unraveling these interrelations. Two important moments in the clinical pathway at which the question of prognosis arises were considered. The first prognostic estimate is required after the diagnostic workup, and the second is required in the post-operative setting, when the histopathology report reveals new information. Table 2 shows the finally retained factors and their HRs in the pre- and post-treatment settings. The

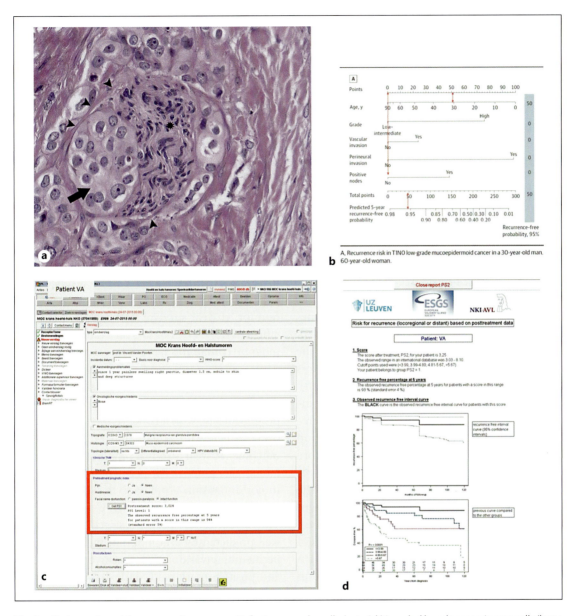

Fig. 1. a Perineural growth as a negative prognostic factor: nerve bundle (asterisk) invaded by adenocarcinoma cells (large arrow) within the perineurium (arrowheads). **b** Nomogram filled out for a 30-year-old patient. **c** Electronic patient file with the prognostic score integrated and automatically calculated. **d** Output of the online software for prognostic score 2.

HR accompanying a given factor, such as perineural growth (fig. 1a), gives the relative risk of recurrence following treatment of a patient who has perineural growth present compared to another who has the baseline level (perineural growth absent), on the condition that the compared patients have equal values for all other variables. Thus, the relative risk of tumor recurrence for a patient with the same set of prognostic factors except for the presence of 'perineural growth' is 2.19 times increased compared to the one without this feature. This HR of 2.19 can be compared to the HR of 2.00 associated with the prognosticator 'skin invasion', and thus, these HRs reflect the relative prognostic impact of 'perineural growth' and 'skin invasion'. Their weighted joint effect can be summarized in a summary prognostic index (table 3) that 'scientifically' estimates the added weighted effect of important factors into 1 number that corresponds to an individualized risk of recurrence for 1 patient. We can now calculate the prognosis of an actual patient based on the combined prognostic effect of features that remained significant in multivariate analysis and, as explained in the next paragraphs, assign the patient to one of four prognostic groups, with a decreasing recurrence-free percentage as the prognostic score increases. In this way, our clinical impression that a patient's prognosis is good, somewhat worse, intermediate, or poor is substantiated.

Based on the First Model in table 2, Prognostic Score 1 Summarizes the Risk Based on Factors Known before Starting Treatment (table 3, left). Multiplying the numeric value for every factor included (e.g. S = 2 for skin invasion) by its corresponding weight (e.g. 0.63) and then adding up the numbers obtained produces prognostic score PS1. This one number reflects the pre-treatment prognostic estimate for this specific patient. The combined effect of all 6 factors on pre-treatment prognosis using PS1 is made visual using 4 cut-off points, or PS1 < 3.85 (PS1 = 1), PS1 = 3.85–4.74 (PS1 = 2), PS1 = 4.75–5.80 (PS1 = 3), and PS1 > 5.80 (PS1 = 4), which divide the study population into 4 subsets of patients, with good, somewhat worse, intermediate or poor prognosis; the respective 5-year recurrence-free rates (with 95% CIs) are 92% (73–98%), 83% (63–92%), 48% (26–67%) and 23% (9–41%).

Based on the Second Model in table 2, Prognostic Score 2 Summarizes the Risk Based on Factors Revealed from the Pathology Report (table 3, right). PS2 is calculated in the same way for the post-operative setting. Cut-off points of PS2 < 3.99 (PS2 = 1), PS2 = 3.99–4.80 (PS2 = 2), PS2 = 4.81–5.67 (PS2 = 3), and PS2 > 5.67 (PS2 = 4) applied to the range of PS2 values of the study population likewise result in 4 groups, with corresponding 5-year recurrence-free rates (with 95% CIs) of 95% (70–99%), 83% (60–93%), 56% (31–75%) and 42% (22–61%).

Creation of a Nomogram for 'All Sites of Major Salivary Gland Carcinoma'

Recently, interesting studies performed in a large institutional series of patients with cancer of 'all major salivary glands' summarized the results of a Cox proportional hazards analysis into a nomogram [16, 17]. One Cox analysis was performed for the outcome '5-year recurrence' [16], and another was performed on the same population for '10-year survival' and '5-year cause-specific survival' [17]. For '5-year recurrence', a nomogram containing age, grade, vascular and perineural invasion and nodal metastasis was selected. When the nomogram was afterwards applied to the same source population, it performed well prognostically, with a concordance (C-) index of 0.85 (see below for explanation) (fig. 1b). For the outcome '10-year survival', the nomogram contained age, grade, cT4, perineural invasion and tumor dimension. Applied to the source population, the nomogram had a high C-index, or 0.81. The '5-year cause-specific survival' proved to be well predicted by a nomogram including grade, perineural invasion, cT4, positive nodal status and positive margins (C-index of 0.856). These nomograms are awaiting the next step in prognostic research, i.e. external validation. One problem that may be faced in this step is the factor 'grade'. As explained in the introduction of this chapter, grading is still quite controversial, and the term 'grade' is used in many different ways. The fact that there is no clear definition of 'grade' in both publications mentioned here may be an obstacle.

Table 3. Prognostic indices PS1 and PS2 in parotid cancer [14]

Variable	Number to enter in the formula
PS1 = 0.024 A + 0.62 P + 0.44 T + 0.45 N + 0.63 S + 0.91 F	
A = Age at diagnosis	number, in years
P = Pain on presentation	1 = no pain
	2 = pain or numbness
T = Clinical T classification*	T1 (<2 cm) = 0
	T2 (2–4 cm) = 1
	T3 (4–6 cm) = 2
	T4 (>6 cm) = 3
N = Clinical N classification	N0 = 0
	N1 = 1
	N2a = 2
	N2b = 3
	N2c = 4
	N3 = 5
S = Skin invasion	1 = no invasion
	2 = invasion
F = Facial nerve dysfunction	1 = intact function
	2 = paresis-paralysis
PS2 = 0.018 A + 0.39 T + 0.34 N + 0.70 S + 0.56 F + 0.78 PG + 0.65 PM	
A = Age at diagnosis	number, in years
T = Clinical T classification	T1 (<2 cm) = 0
	T2 (2–4 cm) = 1
	T3 (4–6 cm) = 2
	T4 (>6 cm) = 3
N = Clinical N classification	N0 = 0
	N1 = 1
	N2a = 2
	N2b = 3
	N2c = 4
	N3 = 5
S = Skin invasion	1 = no invasion
	2 = invasion
F = Facial nerve dysfunction	1 = intact function
	2 = paresis-paralysis
PG = Perineural growth in the resection specimen	1 = no
	2 = yes
PM = Positive surgical margins	1 = no
	2 = yes

Cancer©, copyright 2003, American Cancer Society. * Union Internationale Contre le Cancer 1992 [15].

The Last Step on the Prognostic Ladder: External Validation

The Concept of External Validation

After the creation of a prognostic system such as the abovementioned PS1, PS2 and nomograms, the next step is to check the external validity to support clinical usefulness outside the source population that was used to create the system [18]. This final step has only been taken for the prognostic indices PS1 and PS2. The reasoning behind the requirement of external validation is that the beautiful prognostic discrimination for PS1 and PS2 published in 1999 [13] unfortunately is likely over-estimated because the validation was derived from the population that was used to construct that index. Among statisticians, this well-known feature is called 'over-optimism' [19], or expecting reduced performance when applying a prognostic index in any other population. Generalized prospective use is only legitimate when this decline in prognostic power is limited after quantified application outside the source population, called 'external validation' [18, 19]. This step implies the collection of a new, independent patient population in which both 'statistical' and 'clinical' validity has to be proven. For every patient in this population, the score has to be calculated by filling out the different variables in the summary index. Then, the predictions have to be compared to the actual outcomes [19]. Statistical validity implies using the new data to check whether the index produced remains the best one that can be constructed statistically. Clinical validity is assumed when the index's prediction is accurate enough. To be sure that accuracy is sufficient, we need a correct discrimination (relative ranking of individual risk) and a good calibration (where the estimated recurrence rate is neither too high nor too low) [18]. The clinical usefulness increases every time that an external validation study in yet another population shows that the accuracy is generalizable (reproducible and transportable) to this other population.

External Validation of Prognostic Score 1 and Prognostic Score 2
National and International Validation Samples
The first external validation attempt was conducted using the National Database of The Dutch Head and Neck Oncology Cooperative Group. During the period of 1985–1994, 231 consecutive patients with primarily treated parotid carcinoma were registered by the University Hospitals of Groningen, Leiden, Maastricht/Heerlen, Nijmegen, Rotterdam and Utrecht (median follow-up 52 months) [14]. The second external validation was performed using an international database containing 239 consecutive patients from the Academic Centers of Cologne (n = 101) in Germany and of Leuven (n = 67) and Brussels (n = 61) in Belgium over the period of 1983–2004 (mean follow-up 55 months) [20].

Clinical and Statistical Validation
To address *clinical validation* in both the national and the international validation groups, discrimination was evaluated first. This entailed (1) calculating the indices for every patient, (2) dividing the obtained range of PS1 and PS2 scores according to the source population cut-off points and (3) drawing the corresponding recurrence-free interval curves [13]. Table 4 (upper part for PS1; lower part for PS2) presents the obtained 5-year recurrence-free intervals in the national and the international validation groups compared to the ones in the source population. Table 4 also displays the percentage of over-optimism we initially had in our 1999 study compared to performance in the national and the international datasets. Technically, this percentage is obtained by entering the newly calculated PS in the Cox model. The resulting over-optimism is estimated as 34% for PS1 and 38% for PS2 in the national validation sample (table 4, third column) and as 37% for PS1 and 29% for PS2 in the international validation sample (table 4, second column); statistically, these are very satisfactory percentages. In the second step, concordance measure C was computed for assessment of accuracy, i.e. the relation between patient ordering and PS and the recurrence time [21]. This measure estimates the proportion of all pairs of patients in whom the order of the observed recurrence-free periods concurs with the order of PS1 and PS2. Typically, a decline in concordance measure C compared to its value in the source population is observed when applying a score to a validation sample. In the national validation sample, this decline goes from 0.80 to 0.74 for PS1 and from 0.78 to 0.71 for PS2 (table 4). This decline is comparable to that observed in external validation attempts of other,

Table 4. Recurrence-free percentages at 5 years, over-optimism and concordance for source and validation samples according to the four levels of PS1 and PS2 in parotid cancer [14, 20]

	International[a]	National[b]	NKI/AvL[c]
PS1			
Level 1	94% (44; 5%)	92% (40; 7%)	92% (29; 5%)
Level 2	86% (74; 5%)	70% (56; 9%)	83% (30; 7%)
Level 3	58% (61; 7%)	59% (60; 11%)	48% (31; 11%)
Level 4	42% (56; 7%)	42% (27; 32%)	23% (28; 9%)
Over-optimism	37%	34%	–
Concordance measure C	0.74	0.74	0.80
Dukes' colorectal cancer C	0.74 [24]	0.78 [23]	0.84 [22]
PS2			
Level 1	93% (58; 4%)	90% (21; 10%)	95% (26; 5%)
Level 2	84% (64; 5%)	87% (56; 7%)	83% (25; 8%)
Level 3	61% (44; 8%)	70% (50; 10%)	56% (26; 12%)
Level 4	40% (64; 7%)	40% (44; 16%)	42% (25; 10%)
Over-optimism	29%	38%	
Concordance measure C	0.74	0.71	0.78
Dukes' colorectal cancer C	0.74 [24]	0.78 [23]	0.84 [22]

Values show the recurrence-free percentage with numbers of patients and standard errors in parentheses, unless otherwise indicated.
[a] Validation sample from the Belgian-German Database. [b] Validation sample from the Dutch Head and Neck Oncology Cooperative Group. [c] Original sample from the Netherlands Cancer Institute (NKI/AvL), from which PS1 and PS2 were derived.

well-accepted prognostic systems, such as the Dukes' classification for rectal cancer, where the initial concordance measure declined from 0.84 [22] to 0.78 in an external validation study by Jass et al. [23] and to 0.74 in a subsequent study by Harrison et al. [24]. In our international validation study, a comparable fall in concordance measure C was observed, from 0.80 to 0.74 for PS1 and from 0.78 to 0.74 for PS2 (table 4). In the third step, calibration was visualized by comparing the observed recurrence-free percentages for the different levels of PS in the national (table 4, third column) [14] and the international (table 4, second column) [20] validation samples to the reported percentages for the source sample. We observed adequate calibration, although the expected over-optimism is illustrated by the observation that the four groups in both validation samples have less diverging prognoses, with relatively broader standard error.

To assess *statistical validation*, the null hypothesis tested for rejection states that the original individual weights (as calculated in the source population) accompanying the selected variables as well as the scale of definition of the variables used can be retained. This implies that inclusion of new patient or tumor characteristics will not improve the predictive power. Using the data in the validation populations, no evidence to reject this null hypothesis was produced, and the proportional hazards assumption was also confirmed [14, 20]. The conclusion of the external validation studies was that in the hierarchy of external validity for predictive systems, top-level 5 (multiple independent validations with life-table analyses) is reached [18]. PS1 and PS2 have been proven to be transportable (in geography, in time, in methodology and in follow-up) and generalizable to patients outside the source population [18].

User-Friendly Translation for Use in the Clinic

A downloadable, fillable form is available at http://www.uzleuven.be/parotid and is now also integrated into the electronic patient files of the Head and Neck Oncology Program of the University Hospitals Leuven (fig. 1c). The clinician can introduce a set of prognostic factors for a patient with a previously untreated parotid carcinoma presenting in the clinician's office and obtain the 5-year recurrence-free percentage, the relative risk of recurrence and the according recurrence-free interval curve to answer the patient's question of 'How likely are we to beat this disease?' The answer will be based on not merely on a 'clinical gut feeling' but also on a scientific summary of the weighted combined effect of important multivariate prognostic factors assessed before surgical treatment (PS1) and further refined using elements from surgical pathology in PS2. The patient will rightfully feel relieved when there is a slight, 5% chance of recurrent disease within the next 5 years, as opposed to being worried when hearing that it is about 60% and perhaps more easily accepting that additional therapy is needed in this more aggressively presenting disease.

As an example of use in the clinic, patient V.A. (woman of 31) presents with a 1 cm nodule of the right parotid. There is no pain, no skin invasion, and no palpable neck node, and facial motility is normal. The pre-treatment prognosis can be quantified by entering the data in PS1 into the website or the electronic patient file (fig. 1c), resulting in a PS1 of 2.9, corresponding to a 5-year recurrence-free interval of 94%. For patients in her situation who are treated with surgery alone, only 6% will not be cured at 5 years. In her case, a partial superficial parotidectomy via facelift incision is performed and reveals a completely resected low-grade MEC of 16 mm, pT1, with no vascular or perineural invasion. After clicking the 'after treatment' button on the website (fig. 1d), PS2 is calculated as 3.25, relating to a 5-year recurrence-free interval of 93%. Five years later, there is, as predicted, no evidence of disease, without additional treatment.

Limitations of the Current Prognostic Index System and Future Developments

Apart from being the only externally validated prognostic system, limitations to PS1 and PS2 are the broad CIs of the 5-year recurrence-free estimates and the fact that these are still statistics on a group level, and not individual estimates. These limitations are inherent to a rare disease such as parotid carcinoma and can only be overcome by repeating the same exercise using larger international, multi-institutional databases. The nomograms produced by the Memorial Sloan Kettering Cancer Center team do provide individual estimates but, at this point in time, remain without external validation. The fact that 'histological type' is not included makes PS2 somewhat counterintuitive to use for some clinicians, and especially pathologists. The very high correlation of histological type with other factors that essentially carry the same information with regard to curing the disease explains this. We saw a strong, significant association of high-grade histology with factors incorporated in the model (advanced age, T and N levels, facial nerve dysfunction, perineural growth and positive resection margins), so adopting histological grade as yet another factor in the model did not further statistically contribute to the explanation of the observed variation in outcome. Following publication of our model, other authors also developed and published a prognostic index that did incorporate histology as a dichotomous high-versus-low-grade tumor group in the final model [25]. This model, however, did not include the factors N classification, skin invasion and perineural growth, and the model merely illustrates the interchangeability of strongly correlated factors in a final multivariate model. A recent attempt to validate Carillo's model and to further validate our model in a Brazilian population did not turn out well for either model [26]. Possible reasons for the less-than-optimal performance of our and Carrillo's indices in the South-American population include the long period from 1955 to 2003, over which 175 patients were gathered; the lack of mention of the quality of follow-up, such as the median follow-up duration, dropout, or death from other causes and inadequate follow-up, which could have resulted in failure to detect recurrences, and especially distant metastasis, leading to over-estimation of oncological results and disease-related survival; and the very low rate of distant metastasis as a cause of failure, whereas this was the main cause of failure in our and Carrillo's studies [27].

Future Developments in Prognostic Research

Prognostic research in the near future should focus on incorporating other and better information in a pre-operative prognostic index (the patient factor 'comorbidity' [28]; indices on a histological subtype level; and for the nonparotid subsites, better assessment of the anatomical extent, such as via diffusion-weighted MRI or CT volumetrics [29]), which would clearly refine our prognostic estimate, as well as including molecular biological factors in a post-operative prognostic estimate [1, 5].

References

1 Vander Poorten V, Hunt J, Bradley PJ, Haigentz M Jr, Rinaldo A, Mendenhall WM, et al: Recent trends in the management of minor salivary gland carcinoma. Head Neck 2014;36:444–455.
2 Skalova A, Vanecek T, Sima R, Laco J, Weinreb I, Perez-Ordonez B, et al: Mammary analogue secretory carcinoma of salivary glands, containing the ETV6-NTRK3 fusion gene: a hitherto undescribed salivary gland tumor entity. Am J Surg Pathol 2010;34:599–608.
3 Barnes L, Eveson JW, Reichart P, Sidranski P: Pathology and Genetics of Head and Neck Tumours. World Health Classification of Tumours. Lyon, IARC Press, 2005, p 210.
4 Brandwein MS, Ivanov K, Wallace DI, Hille JJ, Wang B, Fahmy A, et al: Mucoepidermoid carcinoma: a clinicopathologic study of 80 patients with special reference to histological grading. Am J Surg Pathol 2001;25:835–845.
5 Vander Poorten V, Bradley PJ, Takes RP, Rinaldo A, Woolgar JA, Ferlito A: Diagnosis and management of parotid carcinoma with a special focus on recent advances in molecular biology. Head Neck 2012;34:429–440.

6 Skalova A, Sima R, Vanecek T, Muller S, Korabecna M, Nemcova J, et al: Acinic cell carcinoma with high-grade transformation: a report of 9 cases with immunohistochemical study and analysis of TP53 and HER-2/neu genes. Am J Surg Pathol 2009;33:1137–1145.
7 Coca-Pelaz A, Rodrigo JP, Triantafyllou A, Hunt JL, Rinaldo A, Strojan P, et al: Salivary mucoepidermoid carcinoma revisited. Eur Arch Otorhinolaryngol 2015;272:799–819.
8 Coca-Pelaz A, Rodrigo JP, Bradley PJ, Vander Poorten V, Triantafyllou A, Hunt JL, et al: Adenoid cystic carcinoma of the head and neck – an update. Oral Oncol 2015;51:652–661.
9 Gomez DR, Katabi N, Zhung J, Wolden SL, Zelefsky MJ, Kraus DH, et al: Clinical and pathologic prognostic features in acinic cell carcinoma of the parotid gland. Cancer 2009;115:2128–2137.
10 Guntinas-Lichius O, Wendt TG, Buentzel J, Esser D, Boger D, Mueller AH, et al: Incidence, treatment, and outcome of parotid carcinoma, 1996–2011: a population-based study in Thuringia, Germany. J Cancer Res Clin Oncol 2015;141:1679–1688.

11 Kleinbaum DG: Kaplan-Meier survival curves and the log-rank test; in Kleinbaum DG (ed): Survival Analysis: A Self Learning Text, ed 1. New York, Springer, 1996, pp 45–68.
12 Vander Poorten VLM: Salivary Gland Carcinoma: Stepping Up the Prognostic Ladder. Leuven, Acco Publishers, 2002.
13 Vander Poorten VLM, Balm AJM, Hilgers FJM, Tan IB, Loftus-Coll BM, Keus RB, et al: The development of a prognostic score for patients with parotid carcinoma. Cancer 1999;85:2057–2067.
14 Vander Poorten VL, Hart AA, van der Laan BF, Baatenburg de Jong RJ, Manni JJ, Marres HA, et al: Prognostic index for patients with parotid carcinoma: external validation using the nationwide 1985–1994 Dutch Head and Neck Oncology Cooperative Group database. Cancer 2003;97:1453–1463.
15 Sobin LH, Wittekind C (eds): TNM Classification of Malignant Tumours, ed 6. New York, Wiley-Liss, 2002.
16 Ali S, Palmer FL, Yu C, Dilorenzo M, Shah JP, Kattan MW, et al: A predictive nomogram for recurrence of carcinoma of the major salivary glands. JAMA Otolaryngol Head Neck Surg 2013;139:698–705.

17 Ali S, Palmer FL, Yu C, Dilorenzo M, Shah JP, Kattan MW, et al: Postoperative nomograms predictive of survival after surgical management of malignant tumors of the major salivary glands. Ann Surg Oncol 2014;21:637–642.
18 Justice AC, Covinsky KE, Berlin JA: Assessing the generalizability of prognostic information. Ann Intern Med 1999;130:515–524.
19 Altman DG, Royston P: What do we mean by validating a prognostic model? Stat Med 2000;19:453–473.
20 Vander Poorten V, Hart A, Vauterin T, Jeunen G, Schoenaers J, Hamoir M, et al: Prognostic index for patients with parotid carcinoma: international external validation in a Belgian-German database. Cancer 2009;115:540–550.
21 Harrell FE, Lee KL, Mark DB: Tutorial in biostatistics. Multivariable prognostic models: issues in developing models, evaluating assumptions and adequacy, and measuring and reducing errors. Stat Med 1996;15:361–387.
22 Dukes CE, Bussey HJR: The spread of rectal cancer and its effect on prognosis. Br J Cancer 1958;12:309–320.
23 Jass JR, Love SB, Northover JMA: A new prognostic classification of rectal cancer. Lancet 1987;1:1303–1306.
24 Harrison JC, Dean PJ, el Zeky F, Vander Zwaag R: From Dukes through Jass: pathological prognostic indicators in rectal cancer [see comments]. Hum Pathol 1994;25:498–505.
25 Carrillo JF, Vazquez R, Ramirez-Ortega MC, Cano A, Ochoa-Carrillo FJ, Onate-Ocana LF: Multivariate prediction of the probability of recurrence in patients with carcinoma of the parotid gland. Cancer 2007;109:2043–2051.
26 Takahama A Jr, Sanabria A, Benevides GM, de Almeida OP, Kowalski LP: Comparison of two prognostic scores for patients with parotid carcinoma. Head Neck 2009;31:1188–1195.
27 Vander Poorten VLM, Balm A, Hart A: Reflections on Takahama Jr. et al.'s 'Comparison of two prognostic scores for patients with parotid carcinoma'. Head Neck 2010;32:274–275.
28 Terhaard CH, van der Schroeff MP, van SK, Eerenstein SE, Lubsen H, Kaanders JH, et al: The prognostic role of comorbidity in salivary gland carcinoma. Cancer 2008;113:1572–1579.
29 Vandecaveye V, De Keyser F, Vander Poorten, V, Dirix P, Verbeken E, Nuyts S, et al: Head and neck squamous cell carcinoma: value of diffusion-weighted MR imaging for nodal staging. Radiology 2009;251:134–146.
30 Spiro RH: Salivary neoplasms: overview of a 35-year experience with 2,807 patients. Head Neck 1986;8:177–184.
31 Spiro RH, Armstrong J, Harrison L, Geller NL, Lin SY, Strong EW: Carcinoma of major salivary glands. Recent trends. Arch Otolaryngol Head Neck Surg 1989;115:316–321.
32 Kane WJ, McCaffrey TV, Olsen KD, Lewis JE: Primary parotid malignancies. A clinical and pathologic review. Arch Otolaryngol Head Neck Surg 1991;117:307–315.
33 Poulsen MG, Pratt GR, Kynaston B, Tripcony LB: Prognostic variables in malignant epithelial tumors of the parotid. Int J Radiat Oncol Biol Phys 1992;23:327–332.
34 Leverstein H, van der Wal JE, Tiwari RM, Tobi H, van der Waal I, Mehta DM, Snow GB: Malignant epithelial parotid gland tumours: analysis and results in 65 previously untreated patients. Br J Surg 1998;85:1267–1272.
35 Therkildsen MH, Christensen M, Andersen LJ, Schiodt T, Hansen HS: Salivary gland carcinomas – prognostic factors. Acta Oncol 1998;37:701–713.
36 Renehan AG, Gleave EN, Slevin NJ, McGurk M: Clinico-pathological and treatment-related factors influencing survival in parotid cancer. Br J Cancer 1999;80:1296–1300.
37 Harbo G, Bundgaard T, Pedersen D, Sogaard H, Overgaard J: Prognostic indicators for malignant tumours of the parotid gland. Clin Otolaryngol 2002;27:512–516.
38 Godballe C, Schultz JH, Krogdahl A, Moller-Grontved A, Johansen J: Parotid carcinoma: impact of clinical factors on prognosis in a histologically revised series. Laryngoscope 2003;113:1411–1417.
39 Lima RA, Tavares MR, Dias FL, Kligerman J, Nascimento MF, Barbosa MM, Cernea CR, Soares JR, Santos IC, Salviano S: Clinical prognostic factors in malignant parotid gland tumors. Otolaryngol Head Neck Surg 2005;133:702–708.
40 Mendenhall WM, Morris CG, Amdur RJ, Werning JW, Villaret DB: Radiotherapy alone or combined with surgery for salivary gland carcinoma. Cancer 2005;103:2544–2550.

Vincent Vander Poorten, Professor and Deputy Clinical Head
Otorhinolaryngology and Head and Neck Surgery, Section Head and Neck Oncology
Section Head, Department of Oncology, University Hospitals Leuven
Herestraat 49, BE–3000 Leuven (Belgium)
E-Mail vincent.vanderpoorten@uzleuven.be

Surgery for Primary Malignant Parotid Neoplasms

Daniel G. Deschler[a] · David W. Eisele[b]

[a]Massachusetts Eye and Ear Infirmary, Department of Otology and Laryngology, Harvard Medical School, Boston, Mass., and [b]Department of Otolaryngology – Head and Neck Surgery, The Johns Hopkins University School of Medicine, Baltimore, Md., USA

Abstract

The successful treatment of salivary gland malignancies originating in the parotid gland begins with rigorous and thorough surgical management coupled with the directed and appropriate potential adjuvant use of radiation therapy and chemotherapy. The anatomic complexity of the region in relation to the facial nerve and adjoining neurovascular and musculoskeletal structures requires sound surgical planning and decision making based on preoperative and intraoperative findings. The clinical presentation of parotid malignancy is summarized, as well as the further evaluations that are performed, including imaging and tissue biopsy. A systematic approach to ensure resection with clear margins, management of the facial nerve, applicable extensions of the primary resection and management of the neck are presented. The key determinants of disease control, including the stage (early or advanced), histologic grade (low or high) and margin status (clear or positive), are highlighted as critical aspects of surgical management.

© 2016 S. Karger AG, Basel

Introduction

Parotid malignancies constitute a relatively rare form of human cancer. SEER database reviews indicate that salivary gland malignancies have an incidence rate of 0.89–1.2 per 100,000 per year [1, 2]. This rate accounts for less than 5% of all head and neck cancers presenting annually. Of note, greater than one half of all salivary gland malignancies present in the parotid gland [3]. These neoplasms typically occur during the 6th decade of life, but the presentation can vary according to the specific pathology [4]. The SEER data also suggests a slight male predominance, while other studies have indicated that both genders are equally affected [3]. Although equivalent incidence rates for all races have been noted for the common malignant histologic subtypes of mucoepidermoid carcinoma (MEC) and adenoid cystic carcinoma (ACC), most other subtypes have higher incidence rates in whites. There continues to be great debate over the concept that there is a higher rate of malignant histology in salivary gland masses that occur in children [5].

The unique position of the parotid gland within the head and neck region accounts for the specific findings associated with malignant disease of this gland as well as the subsequent management challenges. The gland extends from the zygoma superiorly to the digastric muscle and jugulodigastric region inferiorly. Posteriorly, the gland extends from the auricular cartilages and bony external auditory canal of the temporal bone to the anterior buccal space. Superficially, the gland is located just below the subdermal layer and then extends around the mandibular ramus into the parapharyngeal space. Notably, the facial nerve directly enters the gland upon its exit from the stylomastoid foramen and ramifies throughout the gland as it travels to the muscles of the face. The branches of sensory nerves likewise run throughout the gland, and the continuation of the external carotid artery system travels through the deep gland. The deep extension of the gland into the parapharyngeal space affords proximity to the skull base and the deeply located neurovascular structures.

Presentation

Although most parotid tumors present as a painless mass, specific findings on physical examination may raise the suspicion of malignancy [3]. Fixation of the mass to any of the adjoining structures is highly indicative of a malignant neoplasm. Specific areas to be considered include the skin, tragal cartilage, underlying mandible and muscles of mastication, as well as the deep structures of the upper neck. Pain upon palpation of the mass likewise indicates a potential malignant process, especially when combined with some degree of fixation. Critical assessment of the cranial nerves must be undertaken, with specific attention placed on any dysfunction of the facial nerve. Specific testing of all branches should be performed, as involvement of a specific branch might be quite subtle and present only upon direct testing of the muscular group, or it might manifest as muscle fasciculations in that region. Likewise, because of the dense sensory innervation of the region, the presence of sensory loss in the face, temporal, or scalp region may indicate an advanced malignant neoplasm.

Increasingly, asymptomatic parotid masses are found incidentally on images obtained for other reasons. These masses may be subsequently completely undetectable on physical examination, but thorough evaluation should be undertaken with appropriate concern for malignancy [6].

When pain is the presenting symptom and a mass is identified in the parotid region, a high degree of suspicion of malignancy must be considered, as about one third of malignant tumors will present with pain. Likewise, functional impairment of the facial nerve with either paresis or complete paralysis in the setting of a parotid mass should indicate malignancy, unless proven otherwise. Similarly, new onset of facial paralysis deserves consideration of an underlying parotid mass, especially if it is of prolonged duration. Physical examination of the region should document any seventh cranial nerve deficits, as well as deficits of other cranial nerves in the region, such as V, IX, X, XI and XII. Impairment of jaw movement resulting in trismus or fixation to adjoining structures should likewise be identified, in addition to prominent or suspicious cervical lymphadenopathy.

Evaluation

Upon completion of the history and physical examination, further evaluation of a parotid mass focuses on anatomic definition with appropriate imaging studies and histologic characterization. Although increasing experience is accruing with ultrasound imaging of the parotid area [7, 8], limitations persist in the anatomic detail afforded by ultrasound. Axial imaging obtained with magnetic resonance imaging with gadolinium provides the most benefit in defining the tumor extent and relationship to critical adjoining structures [9, 10]. Magnetic resonance imaging provides superior soft tissue assessment compared to CT scanning, as well as the assessment of potential perineural involvement [11]. CT does enable excellent assessments of bone and regional lymph nodes. Positron emission testing with CT plays a limited role in the assessment of primary parotid neoplasms, but it can be beneficial in evaluation of the presence or extent of metastatic disease, both regional and distal [12, 13]. Although it is not indicated for all parotid malignancies, positron emission testing with CT should be considered for aggressive high-grade diseases, such as salivary ductal carcinoma (SDC), carcinoma ex-pleomorphic adenoma (CXPA), and solid high-grade adenoid cystic carcinoma (ACC).

The benefits of fine-needle aspiration biopsy (FNAB) in the assessment of salivary gland masses have been well established, especially when performed at high-volume, experienced centers [14–16]. Although helpful, FNAB should not be considered definitively diagnostic but more suggestive of a specific pathologic process. FNAB results can reinforce suspicion of an underlying malignancy, but ultimate treatment decisions cannot be uniformly based solely on the results. FNAB can be especially helpful when combined with flow cytometry in the setting of an aggressive parotid process to diagnose a potential lymphoma rather than disease of primary parotid origin [17], and the results have significant impacts on treatment planning.

It is important to note that although FNAB results may be consistent with a neoplasm, the final diagnosis based on histopathology may differ. Masses initially considered benign on FNAB or nondiagnostic specimens can ultimately be found to be malignant. Therefore, in cases of uncertainty, resection is usually indicated.

Surgical Treatment

Although the majority of parotid tumors are benign, 15–20% are malignant. Malignancy may be considered more likely based on specific physical examination findings, on the patient's history, and on the results of imaging studies obtained preoperatively, as well as on histologic information obtained by FNAB. Even if none of the findings of any of these investigations indicate malignancy, a final diagnosis of malignancy is still not an uncommon occurrence. For this reason, one should always have an index of suspicion for a malignant diagnosis and should never be surprised when it is determined to be the final diagnosis.

The central theme that overrides the management of a malignant parotid neoplasm is the achievement of complete surgical resection with appropriate clear margins. The National Comprehensive Cancer Network (NCCN) guidelines for the management of salivary gland malignancies, as outlined in figure 1, emphasize this principle in the recommended treatment for disease of all stages, except for unresectable T4b and advanced metastatic disease [18]. Appropriate treatment of the neck is based on the tumor stage and histology. Adjuvant multidisciplinary treatment is based on the final pathologic stage and tumor grade, as these are the primary determining factors for survival. Figure 2 demonstrates survival by stage [19], and figure 3 shows survival by histologic grade.

In the setting of a fully functional facial nerve preoperatively, preservation of the nerve is warranted. The most effective means of achieving this goal are based on the traditional method of utilizing a standard parotidectomy approach with facial nerve identification, dissection, and preservation. Although traditional terms, such as superficial and total parotidectomy, provide a foundation for describing such procedures, the most important prognostic feature is the achieving of an adequate amount of uninvolved tissue

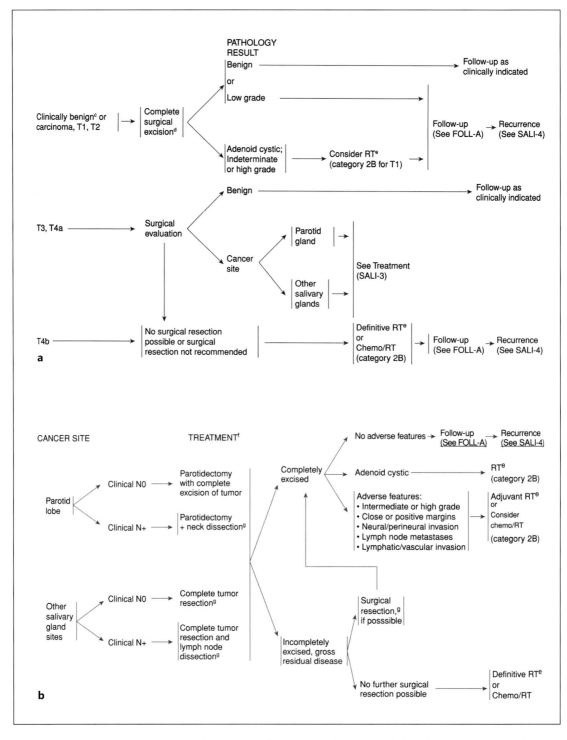

Fig. 1. a NCCN management guidelines for parotid malignancy [18]. **b** NCCN guidelines for management of the neck with parotid malignancy [18].

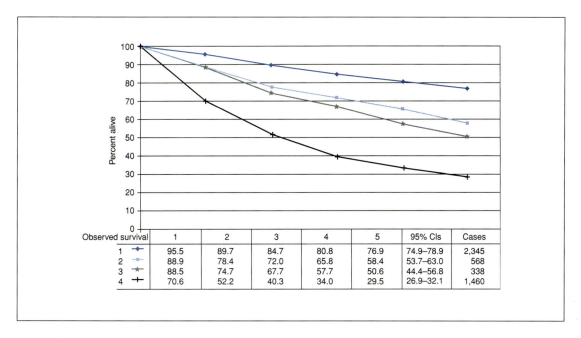

Fig. 2. Observed survival rates for salivary gland malignancies based on stage [19]. Five-year observed survival rates by 'combined' American Joint Committee on Cancer (AJCC) stage for cancer of the major salivary glands, 1998–1999 [95% confidence intervals (CIs) correspond to 5-year survival rates].

Histology	5-year survival
Low-grade mucoepidermoid carcinoma	92–100%
Polymorphous low-grade adeno carcinoma	>90%
Acinic cell carcinoma	85–90%
Intermediate-grade mucoepidermoid carcinoma	70–95%
Adenoid cystic carcinoma	80–90% (60% at 10 yrs) (40% at 20 yrs)
Carcinoma ex-pleomorphic adenoma	30%
High-grade mucoepidermoid carcinoma	22–51%
Adenocarcinoma	10–60%
Salivary duct carcinoma	20%

Fig. 3. Five-year survival rates based on tumor histology (low grade to high grade).

around the tumor in all appropriate directions. If malignancy is expected preoperatively, procedures that may not inherently provide adequate margin control for the majority of the tumor should be avoided. Therefore, procedures such as a tumor enucleation and extracapsular dissection are generally to be avoided [20]. For malignancy that is unexpectedly discovered during or following an extracapsular dissection procedure, revision surgery has been demonstrated to be effective with good oncologic outcomes [21, 22].

In the setting of an advanced neoplasm with increased tumor size and evidence of extra-tumoral extension, significant extension of the scope of resection may be warranted. The most common of these extensions involve resection of specific facial nerve branches or the main trunk of the facial nerve and the involved distal branches. In this context, histological confirmation of malignancy is obtained prior to any resection of the facial nerve.

If the facial nerve function is significantly affected preoperatively and extensive infiltrative disease is noted on physical examination and imaging studies, consideration should be given to radical parotidectomy with facial nerve sacrifice as the initial procedure. Facial nerve branches can be appropriately identified and cleared of tumor involvement distally and proximally prior to nerve grafting. With marked tumor infiltration, other extensions of surgical resection may include the underlying masseter muscle, the abutting auricular cartilage, the temporal bone and the overlying skin. Further resection may also include removal of the ramus of the mandible as well as the condyle and temporal bone [23]. Such procedures often require resection of the external carotid artery system, and deep extension into the parapharyngeal space may likewise require resection of neural and vascular structures.

If significant extensions of the standard parotidectomy procedures are to be considered, then they should be performed with appropriate restraint and consistency to limit additional morbidity. When operating in this region, one is advised to adhere to the basic principle that the closest margin achieved during resection at any interface will be the closest margin on final pathology. Therefore, if one is willing to tolerate a close margin in one anatomic region, such as an area adjacent to the facial nerve, it would not make sense to then engage in significant and potentially morbid extensions in other directions to obtain a markedly wider margin.

Prior to any intervention, the surgical procedure should be discussed with the patient in detail, including the appropriate risks, benefits, sequelae, and potential complications. Similarly, any likely or possible extensions of the procedure, such as facial nerve resection, should be discussed preoperatively so that if they are required, they can be performed with the patient's consent and their family's knowledge.

Resection of a parotid neoplasm in the setting of a fully functional facial nerve begins with appropriate incision planning that affords adequate access to the region and to any possible extensions that may be warranted. For this reason, cosmetic approaches, such as the facelift approach, are generally avoided so as to not limit ultimate access in tumor resection, facial nerve dissection, or dissection of the neck. Facial nerve monitoring is a well-established adjunct to the procedure [24]. The standard parotidectomy incision has been well described, extending from the pretragal crease around the lobule and onto the mastoid and extending forward in a natural neck crease in the upper neck. Such an incision may be extended further anteriorly toward the submentum to allow for access to the level I lymphatics should resection of these nodes be required. If access to the lower cervical lymphatics is required, then consideration should be given to a hemiapron incision that extends down the sternocleidomastoid muscle toward the sternal notch. In addition to providing adequate access to the lower jugular chain and posterior triangle lymphatics, such an incision may be beneficial by allowing for the utilization of a potential cervicofacial advancement flap should it be required in the context of significant skin involvement by the primary tumor. Another option is creation of a McFee-type incision lower in the neck. Upon completion of the incision, the facial flap is developed with the 'deep plane technique,' allowing for elevation just superficial to the parotidomasseteric fascia, with care taken to ensure for an adequate lateral tumor margin. This dissection is performed to the anterior aspect of the parotid gland. The greater auricular nerve is commonly sacrificed in the setting of malignancy, but in select instances, it can be safely preserved. This nerve can be used for later facial nerve reconstructions if needed.

The posterior aspect of the gland is mobilized by separating the fascial attachments to the digastric muscle and the mastoid tip. Further mobilization is completed superiorly by dividing the fascial attachments to the tragus and tympanic bone, thereby exposing the tympanomastoid suture line. Meticulous dissection is then performed through these tissues to identify the main trunk of the facial nerve, at which point further anterior dissection is undertaken, identifying the *pes anserinus* and the main branches of the nerve. Appropriate further dissection of the gland is undertaken to fully mobilize the mass with the surrounding parotid parenchyma. The extent of dissection around the tumor is determined by palpation and visualization of the tissues abutting the tumor. For small well-circumscribed neoplasms, successful complete resection with clear surgical margins can be achieved with partial parotidectomy. If a high-grade histology is suspected or there is evidence of neck metastases, formal superficial parotidectomy should be performed to encompass potential intraparotid lymph node metastases, as intraparotid lymph nodes are highly associated with neck metastases [25, 26]. The deep lobe should be inspected and resected if there is suspicion of deep lobe nodal metastases. Larger masses may also require more complete superficial parotidectomy with mobilization of the entire lateral lobe of the gland in the inferior to superior direction, superior to inferior direction, or a combination of both depending on the tumor location.

Various techniques may be utilized for division of the parotid parenchyma, including cold scalpel, diathermy, thermal scalpel, and harmonic scalpel [27–29]. The choice of the exact technique is dependent upon the user's experience with these instruments for hemostasis and the minimization of neural trauma. Nerve dissection may also be done in a retrograde fashion by identifying peripheral nerve branches, such as the marginal mandibular, buccal, or zygomatic branches, as they exit the gland and then dissecting retrograde toward the main trunk to facilitate tumor removal.

In the case of a deep lobe neoplasm, the superficial lobe of the gland should be mobilized to provide adequate delineation of the facial nerve with access to the tumor, allowing for appropriate mobilization and preservation of the uninvolved facial nerve during tumor resection. Again, if high-grade histology or cervical metastasis is present, consideration should be given to complete resection of the mobilized superficial lobe to assess the intraparotid lymph nodes. Consideration should also be given to sacrifice of deep lobe structures, such as the retromandibular vein, deep and superficial temporal arteries, and external carotid artery.

Once the specimen has been completely removed, it should be oriented and sent to pathology. Frozen section pathology assessment is performed only if it will guide surgical decision making, as it is unlikely to impact the operative plan established with cytology and imaging studies. If a frozen section is requested, the surgeon must keep the request reasonable. Frozen sections are obtained to confirm the presence of a neoplasm if cytology is nondiagnostic or if the cytopathological diagnosis is at odds with the clinical or intraoperative findings, to determine the local and nodal extents of the disease, or to ensure specimen distribution for special studies, for instance, if lymphoma is suspected.

If the tumor is confirmed as a malignancy on frozen section, the tumor grade *(low vs. high)* and histological appearance *(encapsulated vs. infiltrative)* may guide potential further resection in the primary tumor region, decisions to send frozen section margins from the resection bed, and lymph node sampling or elective neck dissection. While frozen section analysis is being completed, verification of hemostasis and documentation of the integrity of the facial nerve is recommended. Upon completion, appropriate cosmetic closure over suction drains is undertaken.

Management of the Facial Nerve

The previous discussion provides a workable framework for addressing a discrete parotid mass in the setting of a functioning facial nerve that is not directly involved by the neoplasm during parotidectomy. If intraoperative findings reveal close approximation of the tumor to the nerve, efforts should be made to mobilize the functioning normal-appearing nerve as much as possible to allow for complete resection without tumor disruption and with facial nerve preservation. If ultimate margins from such a resection reveal microscopic positivity, adequate control can readily be achieved with the addition of an adjuvant treatment, such as radiation therapy or chemo-radiation therapy, in the setting of aggressive histology.

If the nerve is found to be inseparable from the tumor or obviously involved by the tumor, frozen section confirmation of malignancy should be obtained intraoperatively prior to any consideration of facial nerve resection. If nerve resection is undertaken, the proximal and distal margins of the nerve should be marked and cleared by frozen section analysis prior to nerve reconstruction. If only select branches of the facial nerve are involved, then only those branches should be resected and reconstructed to allow for better recovery (fig. 4a, b). If proximal tracing of the main trunk of the facial nerve is required to achieve a clear surgical margin, appropriate mastoidectomy should be undertaken to obtain a clear proximal margin.

Preoperative physical examination findings, such as partial paresis or full paralysis, as well as radiographic findings of extensive infiltrative involvement, may indicate the necessity of facial nerve resection at the outset of the procedure. In such cases, the malignant diagnosis should be confirmed histologically prior to performing nerve resection. An effort to attempt facial nerve dissection can be entertained but should be readily aborted if dissection into the tumor itself is required. The normal-appearing nerve should be identified both proximally, with mastoidectomy if needed, and distally through retrograde dissection. Once achieved, then a relatively straightforward total parotidectomy is undertaken with appropriate extensions, as required by the tumor extent. Appropriate nerve and tissue reconstruction is performed after complete resection and margin clearance.

Surgical Management of the Neck

Neck dissection at the time of primary resection is indicated for all parotid malignancies with evidence of regional disease (N+) [30] at the time of presentation, as defined by examination, imaging studies, and cytology. Comprehensive resection of the involved neck levels as well as those at risk usually warrants modified radical neck dissection or multilevel selective neck dissection. Because of the inherent increase in stage related to the presence of regional metastases, such patients will almost uniformly be considered for adjuvant radiation therapy (fig. 1b).

Management of the clinically negative neck (N0) is a more challenging issue. The literature indicates that occult neck metastases occur at a rate of approximately 15% for N0 parotid malignancies [30, 31]. This rate is higher in some series that include data on intraparotid lymph nodes. The higher-grade histologies, such as a CXPA, SDC, undifferentiated carcinoma, squamous cell carcinoma, high-grade mucoepidermoid and high-grade adenocarcinoma, have a greater propensity for occult metastases [32]. Therefore, treatment of the neck is recommended and may include elective neck dissection (fig. 1b). Adenoid cystic carcinoma has recently been reported to have a higher regional metastatic rate than previously noted [33]. Although postoperative radiation therapy or chemo-radiation may be part of planned care, elective neck dissection may assist with treatment planning, as well as with defining the true extent of disease.

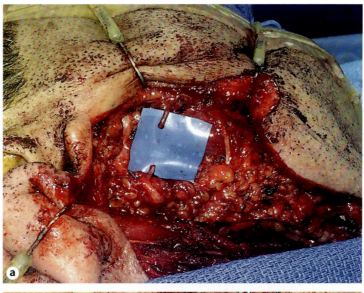
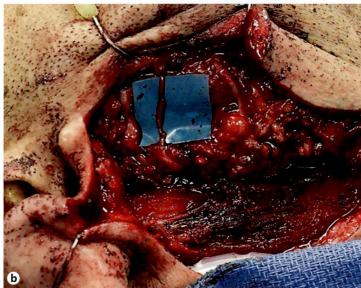

Fig. 4. a Resection of a 3-cm CXPA with parotidectomy and sacrifice of involved buccal branch. **b** Reconstruction of buccal branch after selective research and margin clearance with great auricular nerve branch.

Other factors predictive of occult neck disease include extraparenchymal extension, a tumor size of greater than 4 cm, pain at presentation, facial nerve involvement and the presence of intraparotid positive lymph nodes [34]. As the level II and level III nodal basins are the most likely locations of occult metastases, selective neck dissection of these regions is recommended and readily achieved through a standard parotidectomy incision. If occult disease is identified, appropriate adjuvant treatment can be planned. If the neck is negative on pathologic evaluation and the histology is favorable at the primary site, single modality surgical treatment can be adopted with greater assurance.

Efficacy of Treatment

Using the treatment approach of upfront resection with appropriate neck dissection and surgical extensions combined with adjuvant radiation therapy or chemo-radiation, as determined by the stage and tumor grade, consistent survival results can be obtained. The major determinants of survival remain the tumor stage and grade at presentation. Low-grade tumors have excellent 5-year survival rates: low-grade mucoepidermoid cancer (92–100%) [35–37], acinic cell carcinoma (>90%) [38], and polymorphous low-grade adenocarcinoma (>90%) [39, 40]. Small series have indicated that the recently described mammary analog secretory carcinoma likewise has a similar favorable survival rate [41].

High-grade tumors present more dismal pictures with significantly poorer 5-year survival rates: high-grade MEC (22–42%) [42–44], solid-type ACC (17%) [45–47], high-grade adenocarcinoma (10–60%) [1, 48], SDC (20%) [49, 50], and CXPA/mixed malignant tumor (30%) [51]. Treatment failure can occur due to local, regional or distal recurrence. A special note should be made of the propensity for late distal recurrence, and appropriate long-term follow-up should be maintained for all patients with parotid gland cancer [52].

Conclusions

Primary parotid malignancies remain a relatively rare but treatable form of head and neck cancer. Successful management requires appropriate understanding of the effects of tumor stage and histology grade and the critical role of comprehensive surgical intervention based on thorough preoperative evaluation and insightful intraoperative decision making.

References

1 Wahlberg P, Anderson H, Biörklund A, Möller T, Perfekt R: Carcinoma of the parotid and submandibular glands – a study of survival in 2,465 patients. Oral Oncol 2002;38:706–713.
2 Boukheris H, Curtis RE, Land CE, Dores GM: Incidence of carcinoma of the major salivary glands according to the WHO classification, 1992 to 2006: a population-based study in the United States. Cancer Epidemiol Biomark Prev 2009;18:2899–2906.
3 Bjørndal K, Krogdahl A, Therkildsen MH, Overgaard J, Johansen J, Kristensen CA, et al: Salivary gland carcinoma in Denmark 1990–2005: a national study of incidence, site and histology. Results of the Danish Head and Neck Cancer Group (DAHANCA). Oral Oncol 2011; 47:677–682.
4 Pinkston JA, Cole P: Incidence rates of salivary gland tumors: results from a population-based study. Otolaryngol Head Neck Surg 1999;120:834–840.
5 Stevens E, Andreasen S, Bjørndal K, Homøe P: Tumors in the parotid are not relatively more often malignant in children than in adults. Int J Pediatr Otorhinolaryngol 2015;79:1192–1195.
6 Seo YL, Yoon DY, Baek S, Lim KJ, Yun EJ, Cho YK, et al: Incidental focal FDG uptake in the parotid glands on PET/CT in patients with head and neck malignancy. Eur Radiol 2015;25:171–177.
7 Badea AF, Bran S, Tamas-Szora A, Floareș A, Badea R, Baciut G: Solid parotid tumors: an individual and integrative analysis of various ultrasonographic criteria. A prospective and observational study. Med Ultrason 2013;15:289–298.
8 Yonetsu K, Ohki M, Kumazawa S, Eida S, Sumi M, Nakamura T: Parotid tumors: differentiation of benign and malignant tumors with quantitative sonographic analyses. Ultrasound Med Biol 2004;30:567–574.
9 Kinoshita T, Ishii K, Naganuma H, Okitsu T: MR imaging findings of parotid tumors with pathologic diagnostic clues: a pictorial essay. Clin Imaging 2004;28:93–101.
10 Okahara M, Kiyosue H, Hori Y, Matsumoto A, Mori H, Yokoyama S: Parotid tumors: MR imaging with pathological correlation. Eur Radiol 2003;13(suppl 4): L25–L33.
11 Koyuncu M, Seşen T, Akan H, Ismailoglu AA, Tanyeri Y, Tekat A, et al: Comparison of computed tomography and magnetic resonance imaging in the diagnosis of parotid tumors. Otolaryngol Head Neck Surg 2003;129:726–732.
12 Razfar A, Heron DE, Branstetter BF, Seethala RR, Ferris RL: Positron emission tomography-computed tomography adds to the management of salivary gland malignancies. Laryngoscope 2010; 120:734–738.

13 Kim M-J, Kim JS, Roh J-L, Lee JH, Cho K-J, Choi S-H, et al: Utility of 18F-FDG PET/CT for detecting neck metastasis in patients with salivary gland carcinomas: preoperative planning for necessity and extent of neck dissection. Ann Surg Oncol 2013;20:899–905.

14 Postema RJ, van Velthuysen M-LF, van den Brekel MWM, Balm AJM, Peterse JL: Accuracy of fine-needle aspiration cytology of salivary gland lesions in The Netherlands cancer institute. Head Neck 2004;26:418–424.

15 Cohen EG, Patel SG, Lin O, Boyle JO, Kraus DH, Singh B, et al: Fine-needle aspiration biopsy of salivary gland lesions in a selected patient population. Arch Otolaryngol Head Neck Surg 2004; 130:773–778.

16 Seethala RR, LiVolsi VA, Baloch ZW: Relative accuracy of fine-needle aspiration and frozen section in the diagnosis of lesions of the parotid gland. Head Neck 2005;27:217–223.

17 MacCallum PL, Lampe HB, Cramer H, Matthews TW: Fine-needle aspiration cytology of lymphoid lesions of the salivary gland: a review of 35 cases. J Otolaryngol 1996;25:300–304.

18 Pfister DG, et al: Head and Neck Cancers, Version 1.2015. J Natl Compr Canc Netw 2015;7:847–855.

19 AJCC: Cancer Staging Manual, ed 7. New York, Springer, 2010, p 81.

20 Deschler DG: Extracapsular dissection of benign parotid tumors. JAMA Otolaryngol Head Neck Surg 2014;140:770–771.

21 Mantsopoulos K, Velegrakis S, Iro H: Unexpected detection of parotid gland malignancy during primary extracapsular dissection. Otolaryngol Head Neck Surg 2015;152:1042–1047.

22 Ryu IS, Roh J-L, Cho K-J, Lee S, Choi S-H, Nam SY, et al: Clinical outcomes of patients with salivary gland carcinomas preoperatively misdiagnosed as benign lesions. Head Neck 2013;35:1764–1770.

23 Gidley PW, Thompson CR, Roberts DB, Weber RS: The results of temporal bone surgery for advanced or recurrent tumors of the parotid gland. Laryngoscope 2011;121:1702–1707.

24 Sood AJ, Houlton JJ, Nguyen SA, Gillespie MB: Facial nerve monitoring during parotidectomy: a systematic review and meta-analysis. Otolaryngol Head Neck Surg 2015;152:631–637.

25 Lim CM, Gilbert MR, Johnson JT, Kim S: Clinical significance of intraparotid lymph node metastasis in primary parotid cancer. Head Neck 2014;36:1634–1637.

26 Wang Y-L, Li D-S, Gan H-L, Lu Z-W, Li H, Zhu G-P, et al: Predictive index for lymph node management of major salivary gland cancer. Laryngoscope 2012; 122:1497–1506.

27 Blankenship DR, Gourin CG, Porubsky EA, Porubsky ES, Klippert FN, Whitaker EG, et al: Harmonic scalpel versus cold knife dissection in superficial parotidectomy. Otolaryngol Head Neck Surg 2004;131:397–400.

28 Deganello A, Meccariello G, Busoni M, Parrinello G, Bertolai R, Gallo O: Dissection with harmonic scalpel versus cold instruments in parotid surgery. B-ENT 2014;10:175–178.

29 Ramadan HH, Wax MA, Itani M: The Shaw scalpel and development of facial nerve paresis after superficial parotidectomy. Arch Otolaryngol Head Neck Surg 1998;124:296–298.

30 Armstrong JG, Harrison LB, Thaler HT, Friedlander-Klar H, Fass DE, Zelefsky MJ, et al: The indications for elective treatment of the neck in cancer of the major salivary glands. Cancer 1992;69: 615–619.

31 Lau VH, Aouad R, Farwell DG, Donald PJ, Chen AM: Patterns of nodal involvement for clinically N0 salivary gland carcinoma: refining the role of elective neck irradiation. Head Neck 2014;36:1435–1439.

32 Stennert E, Kisner D, Jungehuelsing M, Guntinas-Lichius O, Schröder U, Eckel HE, et al: High incidence of lymph node metastasis in major salivary gland cancer. Arch Otolaryngol Head Neck Surg 2003;129:720–723.

33 Amit M, Binenbaum Y, Sharma K, Ramer N, Ramer I, Agbetoba A, et al: Incidence of cervical lymph node metastasis and its association with outcomes in patients with adenoid cystic carcinoma. Head Neck 2015;37:1032–1037.

34 Kelley DJ, Spiro RH: Management of the neck in parotid carcinoma. Am J Surg 1996;172:695–697.

35 Goode RK, Auclair PL, Ellis GL: Mucoepidermoid carcinoma of the major salivary glands: clinical and histopathologic analysis of 234 cases with evaluation of grading criteria. Cancer 1998;82:1217–1224.

36 Brandwein MS, Ivanov K, Wallace DI, Hille JJ, Wang B, Fahmy A, et al: Mucoepidermoid carcinoma: a clinicopathologic study of 80 patients with special reference to histological grading. Am J Surg Pathol 2001;25:835–845.

37 Richter SM, Friedmann P, Mourad WF, Hu KS, Persky MS, Harrison LB: Postoperative radiation therapy for small, low-/intermediate-grade parotid tumors with close and/or positive surgical margins. Head Neck 2012;34: 953–955.

38 Hoffman HT, Karnell LH, Robinson RA, Pinkston JA, Menck HR: National Cancer Data Base report on cancer of the head and neck: acinic cell carcinoma. Head Neck 1999;21:297–309.

39 Castle JT, Thompson LD, Frommelt RA, Wenig BM, Kessler HP: Polymorphous low grade adenocarcinoma: a clinicopathologic study of 164 cases. Cancer 1999;86:207–219.

40 Evans HL, Luna MA: Polymorphous low-grade adenocarcinoma: a study of 40 cases with long-term follow up and an evaluation of the importance of papillary areas. Am J Surg Pathol 2000;24: 1319–1328.

41 Sethi R, Kozin E, Remenschneider A, Meier J, VanderLaan P, Faquin W, et al: Mammary analogue secretory carcinoma: update on a new diagnosis of salivary gland malignancy. Laryngoscope 2014;124:188–195.

42 Spiro RH, Huvos AG, Berk R, Strong EW: Mucoepidermoid carcinoma of salivary gland origin. A clinicopathologic study of 367 cases. Am J Surg 1978; 136:461–468.

43 McHugh CH, Roberts DB, El-Naggar AK, Hanna EY, Garden AS, Kies MS, et al: Prognostic factors in mucoepidermoid carcinoma of the salivary glands. Cancer 2012;118:3928–3936.

44 Emerick KS, Fabian RL, Deschler DG: Clinical presentation, management, and outcome of high-grade mucoepidermoid carcinoma of the parotid gland. Otolaryngol Head Neck Surg 2007;136:783–787.

45 Da Cruz Perez DE, de Abreu Alves F, Nobuko Nishimoto I, de Almeida OP, Kowalski LP: Prognostic factors in head and neck adenoid cystic carcinoma. Oral Oncol 2006;42:139–146.

46 Khafif A, Anavi Y, Haviv J, Fienmesser R, Calderon S, Marshak G: Adenoid cystic carcinoma of the salivary glands: a 20-year review with long-term follow-up. Ear Nose Throat J 2005;84:662, 664–667.
47 Gurney TA, Eisele DW, Weinberg V, Shin E, Lee N: Adenoid cystic carcinoma of the major salivary glands treated with surgery and radiation. Laryngoscope 2005;115:1278–1282.
48 Li J, Wang BY, Nelson M, Li L, Hu Y, Urken ML, et al: Salivary adenocarcinoma, not otherwise specified: a collection of orphans. Arch Pathol Lab Med 2004;128:1385–1394.
49 Guzzo M, Di Palma S, Grandi C, Molinari R: Salivary duct carcinoma: clinical characteristics and treatment strategies. Head Neck 1997;19:126–133.
50 Hosal AS, Fan C, Barnes L, Myers EN: Salivary duct carcinoma. Otolaryngol Head Neck Surg 2003;129:720–725.
51 Olsen KD, Lewis JE: Carcinoma ex pleomorphic adenoma: a clinicopathologic review. Head Neck 2001;23:705–712.
52 Chen AM, Garcia J, Granchi PJ, Johnson J, Eisele DW: Late recurrence from salivary gland cancer: when does 'cure' mean cure? Cancer 2008;112:340–344.

David W. Eisele, MD, FACS
Department of Otolaryngology – Head and Neck Surgery
The Johns Hopkins University School of Medicine
601 N. Caroline St., Suite 6210
Baltimore, MD 21287-0910 (USA)
E-Mail deisele1@jhmi.edu

Metastatic Cancer to the Parotid

Jonathan Clark[a, b] · Steven Wang[c]

[a]Department of Head and Neck Surgery, Chris O'Brien Lifehouse, Sidney, N.S.W., and [b]Central Clinical School, The University of Sydney, Sydney, N.S.W., Australia; [c]Department of Otolaryngology – Head and Neck Surgery, University of California, San Francisco, Calif., USA

Abstract

In many regions of the world, the most common type of parotid malignancy is metastatic malignancy. Metastatic cutaneous squamous cell carcinoma and metastatic melanoma are the most common pathologies that metastasize to the parotid gland and cervical lymph nodes. Other more rare metastatic malignancies from cutaneous primary sites include Merkel cell carcinoma, pleomorphic sarcoma (previously called malignant fibrous histiocytoma) and metastatic basal cell carcinoma. Systemic metastases can also occur from tumors such as renal cell carcinoma. This review will focus on the management of the two most common causes of parotid malignancy.

© 2016 S. Karger AG, Basel

Advances in the Management of Metastatic Cutaneous Squamous Cell Carcinoma of the Parotid Gland

Introduction

The incidence of nonmelanoma skin cancer (NMSC) continues to rise and has reached epidemic proportions in countries with close proximity to the equator and in individuals with fair skin. The highest incidence has been observed in Northern Queensland, Australia, where the annual age-adjusted incidence of cutaneous squamous cell carcinoma (CSCC) exceeds 805/100,000 in males [1–3]. Fortunately, NMSC is an unlikely cause of death, with cure rates exceeding 95% [4, 5], although in elderly patients, the mortality rate is similar to that of melanoma for that age group, approaching 80 per 100,000 per year in patients over 85 years of age from 2010 to 2012 [3].

Risk factors for NMSC include skin photosensitivity and melanin content, as determined by the Fitzpatrick classification of skin phenotypes, male gender and increased age. Two factors demand special mention, solar exposure and immunodeficiency.

Solar radiation is the single most important etiologic factor. Ultraviolet (UV) B (wavelength 290–320 nm), and to a lesser extent, UVA (320–400 nm), are capable of inducing reversible and irreversible DNA damage to epidermal cells involving tumor suppressor genes (e.g. p53) or proto-oncogenes (e.g. Ras and Raf). Solar radiation

Table 1. Probability of developing nodal metastases based on individual clinicopathological risk factors

Risk factor	Metastatic likelihood, %
Size >2 cm	20–30
Invasion into subcutaneous fat (depth ≥5 mm)	16–45
Poorly differentiated/metatypical/morpheaform	12–32
High-grade histology or desmoplasia	12
Perineural invasion	40–47
Lymphovascular invasion	40
Location of ear or lip	10–30
Local recurrence	25–62
SCC in pre-existing scar (from burn or trauma)	38
Immunosuppression	13–20

also drives clonal expansion of immortalized cells by triggering apoptosis of surrounding UV-damaged cells with wild-type p53, and it adversely affects cell-mediated immunity in the skin by impairing the function of antigen-presenting cells in the skin [6]. Patients with solid organ transplantation have a 100-fold increase in the incidence of skin cancers, proportional to the duration and degree of immunosuppression. Of note, the ratio of basal cell carcinoma to CSCC is reversed in transplant patients, with more patients developing CSCC.

High-Risk Squamous Cell Carcinoma and Predicting Nodal Metastases
Prospective data regarding disease-related outcomes of all patients with CSCC has indicated that the risks of local recurrence, regional metastasis, distant metastasis, and disease-specific death are 5, 5, 1, and 1%, respectively [4]. Multivariable analysis has revealed that increased tumor thickness (≥2 mm), tumor size (≥2 cm), location on the ear and immunosuppression are statistically significant risk factors for metastasis. The rate of nodal metastases has been frequently reported to exceed 20% in patients with high-risk primary features, as shown in table 1 [7, 8]. Most of these data are based on retrospective studies and tend to overestimate the effects of each particular risk factor. The largest prospective study of high-risk primary CSCC was an interim analysis of the sentinel node in CSCC (SNIC) study [9]. In this study, the risk of metastases in high-risk patients was 14%, and this finding is supported by two recent systematic reviews of other sentinel node studies [10, 11]. In the SNIC study, the significant predictors of metastasis were the number of high-risk factors, perineural invasion and lymphovascular invasion. All patients with nodal metastases had thick tumors (>5 mm). Although very few patients enrolled were immunosuppressed, there remains strong data that immunosuppression increases the risk of metastasis, even in otherwise low-risk tumors [12].

Prognosis in Metastatic Cutaneous Squamous Cell Carcinoma and Nodal Staging of Parotid Metastases
Early versions of the American Joint Committee on Cancer (AJCC) TNM staging system did not discriminate between the number, size and location of nodes. A number of studies have attempted to address this problem. The latest (7th) edition (2010) of the AJCC staging manual has made substantial changes by adopting the same nodal classification as that for mucosal head and neck cancer, without providing any real evidence to validate these changes [13]. A number of high-

risk primary factors were incorporated into the T category to increase the stage from I–II for tumors <2 cm in diameter, including >2 mm thickness, Clark's level IV, perineural invasion, location on the ear or nonhair bearing lip, and poorly or undifferentiated histologic grade [14].

O'Brien proposed separating parotid from cervical node involvement using the P/N system [15]. This system was evaluated in a multi-institutional international trial that concluded that advanced P stage (P3) and neck disease (N1/2) were associated with reduced survival [16]. This was the first major step toward recognizing the need for change, but the P/N system was complex and did not stratify risk well within the P and N groups. The N1S3 system combined parotid and cervical metastases in an attempt to simplify nodal staging. N1S3 refers to the number (one or more) and size(s) of nodes (<3 or >3 cm) and has significant predictive capacity for loco-regional control, disease-specific survival and overall survival [17, 18]. The importance of other prognostic factors also needs to be emphasized, such as extracapsular spread (ECS) and soft tissue deposits, one of which is present in 70% of CSCC nodal metastases. The ITEM (Immunosuppression, Treatment, Extranodal spread and Margin of resection) prognostic score separates patients into three risk groups based on these four prognostic variables [19]. The 5-year risk of dying from disease for high-risk patients is 56%, that for moderate-risk patients is 24%, and that for low-risk patients is 6%. Soft-tissue (non-nodal) deposits of CSCC are associated with an even worse prognosis than ECS [20].

Surgical Management of Nodal Metastases to the Parotid Gland

The most common site of nodal metastasis from CSCC is the parotid gland (75%), followed by the level II nodes (40%). The region of the external jugular lymph node adjacent to the tail of the parotid is critical [21]. Although lymphatic drainage from cutaneous tumors is unpredictable, in patients with parotid metastases, there is considerable evidence that the location of the metastases is predictive of further potential sites of nodal disease [22, 23]. Routine preoperative investigations of patients with suspected parotid nodal metastases include ultrasound guided fine-needle aspiration and CT scanning of the parotid gland, neck and chest. MRI is used for patients with suspected perineural invasion or parapharyngeal extension. The role of positron emission tomography (PET) has not been clearly defined; however, in CSCC, distant metastases outside of the chest are uncommon, making it a low-yield investigation.

There are a number of reasonable strategies for addressing the neck in patients with metastatic CSCC to the parotid gland without clinical neck metastases (P1N0). The options include: (1) parotidectomy + selective neck dissection +/– radiotherapy and (2) parotidectomy + radiotherapy of the parotid bed and neck.

Each approach has its relative merits. Levels II and III are crucial for determining whether the neck is involved by metastases. However, for primary tumors posterior to the ear, level V must also be included. Isolated level I involvement only occurs in patients with anterior facial primaries (glabella, nose, lip, and chin). Therefore, in the majority of patients, the combination of parotidectomy and selective levels II/III neck dissection will allow for accurate staging of the neck. Among patients with posterior scalp primaries, 15% have level IV/V metastases; hence, in this group, dissection of levels II–V is recommended [23]. It is important to note that the external jugular node is considered part of the parotid in this context and always needs to be removed.

A cogent argument can be made that neck dissection is unnecessary given that adjuvant radiotherapy is routine for metastatic CSCC [24–26]. However, by demonstrating that the neck is negative, the radiation fields may be limited to the parotid bed alone, thereby reducing the toxicity of the treatment. Furthermore, among select low-risk cases (N1S3 stage I), radiotherapy can be omitted for those with a single parotid node, negative neck, clear margins and no ECS [27]. There is a need for treatment intensification in patients with high-risk disease (N1S3 stage III), for whom disease-specific survival is only 42% at 5 years. Other high-risk groups include immunosuppressed patients and those with positive margins or soft tissue (non-nodal) deposits of CSCC. Unfortunately, there is no evidence to support treatment intensification with chemotherapy at present. The role of postoperative chemoradiation using concurrent carboplatin is being investigated by the Trans-Tasman Radiation Oncology Group in the POST trial [28].

Advances in the Management of Melanoma Metastatic to the Parotid Gland

Introduction

Cutaneous melanoma is the 6th leading cause of cancer among men and women [29, 30]. According to estimates of the American Cancer Society, there will be about 73,870 new melanoma cases and about 9,940 people will die from melanoma in the US in 2015 [29]. Additionally, the incidence of melanoma is on the rise, increasing by 600% over the past 50 years (SEER data) [30]. The rate of increase in the melanoma incidence is greater than that of any other cancer. The lifetime risk of developing melanoma for an individual born in 1935 is approximately 1 in 1,500 [29]. The lifetime risk of melanoma increases to 1 in 250 for an individual born in 1980 and has been estimated to be 1 in 7 for an individual born in 2000 [29]. A disproportionate share of cutaneous melanoma occurs in the head and neck, which represents 8% of the total body surface area but accounts for 25–30% of all melanomas [31]. In addition, melanomas occurring in the head and neck have been reported to be more biologically aggressive, particularly those arising in the scalp. Among the common head and neck sites of melanoma are the face (40–60%), scalp (14–49%), neck (20–29%), and ear (8–11%) [31].

Risk factors for cutaneous melanoma include a history of sun exposure, particularly a history of blistering sunburns at a young age. Fifty percent of melanomas arise in pre-existing nevi, and individuals with more than 20 nevi on their body have a 3-fold increased relative risk of developing melanoma [31]. Other risk factors include fair skin, red hair, freckling, and a family history.

Certain populations may be at an increased risk of melanoma, including those with certain melanoma susceptibility genes [32]. CDKN2a (p17INK4a), the most common melanoma susceptibility gene, is a highly penetrant autosomal dominant gene that accounts for approximately 20–50% of familial melanoma cases [32]. Mutations in the tumor suppressor BAP1 have been reported to enhance the metastatic potential of uveal melanoma and to predispose individuals to both cutaneous and ocular melanoma [32].

Management of Cutaneous Melanoma of the Head and Neck: Predicting Nodal Metastases

The status of the regional lymph nodes is the most important prognostic factor for recurrence and survival in cutaneous melanoma [33]. The primary draining nodal basins depend on the primary disease site. For head and neck cutaneous melanoma, metastases typically travel through superficial lymphatic pathways to the parotid, facial, postauricular, and occipital nodes and the nodes superficial to the sternocleidomastoid, particularly the external jugular node [34, 35]. Connections between superficial lymphatics and deep lymphatics, such as the submandibular, internal jugular chain, and posterior triangle nodes, are unpredictable, and metastatic melanoma may spread first to either superficial or deep nodal basins of the neck. The intraparotid nodes are commonly the first-echelon nodal drainage basins for cutaneous melanomas of the anterior scalp, temple and forehead, and pinna.

The most predictive factor for nodal metastasis in melanoma is tumor thickness, highlighting the importance of full-thickness tissue sampling during the initial biopsy. Melanomas of less than 1 mm in thickness have a low likelihood (<5%) of nodal metastasis, whereas the risk of nodal metastasis increases with a greater depth of invasion (1–2 mm: 15–20% risk; 2–4 mm: 25% risk; and >4 mm: 35% risk) [33]. Other factors that increase the likelihood of metastasis include the presence of ulceration and a mitotic index of ≥1 per mm^2.

Imaging is not routinely performed for patients with thin melanoma without evidence of metastasis on clinical examination. Some groups, including ours, do obtain imaging for thicker melanomas (>4 mm), especially those of scalp origin, as studies have shown a higher incidence of regional or distant metastasis in such cases. Whole-body PET/CT, while expensive, is probably the best imaging modality for assessment of regional and distant metastatic melanoma. Other noninvasive imaging modalities that have been described for the assessment of metastatic melanoma include ultrasound and conventional CT [36]. Ultrasound may detect micrometastases of 4 mm in thickness, but for smaller-volume metastases, ultrasound is inferior to sentinel node biopsy [37]. When there is suspicion, by palpation or imaging, of nodal metastasis in the parotid or neck, fine-needle aspiration biopsy with or without image guidance should be performed.

There is no role for elective node dissection in cutaneous melanoma [38–40]. Cervical or parotid nodes with occult disease may well be missed because of the complexity of the cutaneous lymphatic drainage patterns in the head and neck, which makes determination of first-echelon draining lymphatics unpredictable. Multiple studies have reported that elective neck dissection does not provide survival benefit [38–40].

Sentinel lymph node biopsy is utilized for the assessment of regional nodal status for cutaneous melanoma and has become the standard practice in many countries for the surgical management of stage Ib or higher cutaneous melanoma [41–44]. Detection of occult regional metastases is crucial for determining prognosis, and sentinel node biopsy is the most sensitive and specific modality for regional staging [44]. The benefits of sentinel node biopsy include the following: it provides the most accurate staging of the regional nodes, allows for the selection of patients who may benefit from adjuvant therapy, and provides reassurance to patients who are sentinel node-negative. In addition, it is often required as a screening tool to select appropriate candidates for clinical trials. However, the therapeutic benefit of sentinel node biopsy for melanoma remains controversial. The Multicenter Selective Lymphadenectomy Trial (MSLT-I) found that stage III patients identified through sentinel lymph node biopsy had improved survival compared to patients who developed palpable metastasis under a watchful waiting policy [41]. On the other hand, other studies have failed to show a survival advantage, leading some to question whether sentinel node biopsy provides any therapeutic benefit [45].

Sentinel node biopsy is a demanding technique, perhaps nowhere else more so than in the head and neck, due to the complexity of the lymphatic drainage in this anatomic location. Neck and intraparotid sentinel lymph nodes are often tiny and difficult to access, there is commonly close proximity of the nodal basins to the primary sites, obscuring interpretation of lymphoscintigraphy, and sentinel lymph node biopsy in the head/neck region carries an inherent risk of permanent cranial nerve deficits (facial and spinal accessory nerves) although around 85% of cases show negative results for cancer. These issues are best highlighted in cases in which the sentinel nodes are within the parotid gland. However, multiple studies have attested to the safety and reliability of intraparotid sentinel node biopsy [46–48]. Intraparotid sentinel nodes are most commonly found in the tail of parotid region or preauricular parotid gland superficial to the facial nerve branches. Facial nerve monitoring is often useful for intraparotid sentinel node biopsy. The technique of sentinel node biopsy continues to improve, and advances such as single photon emission computed tomography combined with CT have allowed for 3-dimensional localization of sentinel node mapping, improving the accuracy and efficiency of sentinel node biopsy [49].

Management after detection of positive sentinel nodes is also controversial. In the MSLT-1 trial, completion lymphadenectomy was performed in all cases with a positive sentinel node biopsy [41]. Approximately 13.7% of completion lymphadenectomy specimens showed evidence of additional metastatic nodes, arguing for the necessity of additional removal of lymph nodes in the setting of the discovery of micrometastatic disease. However, given the morbidity of completion lymphadenectomy, some have questioned the necessity of this additional procedure in all cases of positive sentinel nodes, especially when the initial micrometastatic nodal disease is of very low volume [45]. The MSLT-2 trial, which is currently still in progress, is seeking to address this question, comparing completion lymphadectomy for low-volume metastatic sentinel nodes versus close clinical observation, including ultrasound surveillance [50].

Surgical Management of Nodal Metastasis to the Parotid Gland
Patients with nodal metastasis to the parotid gland, whether identified by sentinel node biopsy or with macrometastases, are consequently upstaged to at least stage III disease. In this setting, imaging is indicated to search for further sites of regional disease and to determine whether distant metastasis is also present, which would indicate stage IV disease [51]. Whole-body PET/CT is the best imaging test to perform in this setting, but neck/chest/abdomen CT may also be an adequate screening assessment method. Brain MRI to rule out intracranial metastasis may also be indicated.

Surgical management of patients with melanoma metastatic to the parotid gland includes therapeutic parotidectomy. Parotidectomy performed after a positive intraparotid sentinel node biopsy has been shown to be safe and without increased complications. Facial nerve monitoring is utilized when appropriate. Most commonly, superficial parotidectomy is performed. However, total parotidectomy to remove at-risk deep lobe intraparotid nodes has been advocated [52].

Elective Neck Management for Melanoma Metastatic to the Parotid
In contrast with CSCC, the role of adjuvant radiotherapy in the elective treatment of lymph nodes in melanoma is limited, and surgery is the treatment of choice for at-risk nodes. Surgery is directed at all likely involved nodes, and thus, cervical lymphadenectomy tends to be more comprehensive for melanoma compared to CSCC [35, 53]. However, selective options may be appropriate, for example, the omission of level I when the primary tumor is located in the posterior scalp or neck and the omission of level V when the anterior face is the primary location. Nodes outside of the named levels should be included on an anatomical basis, for example, the facial, external jugular, postauricular, and occipital nodes, according to the primary site. The jugular vein, spinal accessory nerve, and submandibular gland can frequently be preserved, but it is often preferable to remove the sternocleidomastoid muscle en bloc with the nodal specimen to ensure for thorough resection of all at-risk deep and superficial lymphatics.

When there is N+ disease, elective parotidectomy may be appropriate, depending on the site of the cutaneous primary lesion and whether it is likely that draining lymphatics pass through the parotid nodes. For lesions of the scalp, pinna, and posterior face with established nodal metastasis in the neck, it is recommended that superficial parotidectomy should be performed at the time of neck dissection.

Adjuvant Therapy for Melanoma with Metastasis to the Parotid

Prognosis for any stage III melanoma is poor, with more than half of patients suffering from disease relapse within 10 years [33]. Thus, patients with melanoma metastatic to the parotid gland have a guarded prognosis. Recurrence may affect the regional nodes, or more commonly, there may be progression to distant metastatic stage IV disease. Unfortunately, options for adjuvant treatment after completion of surgical treatment for stage III disease are limited in number and efficacy.

Melanoma has traditionally been viewed as a radioresistant cancer, but radiation therapy does have a role in regional control and is offered postoperatively when there is greater than N1 disease or evidence of nodal ECS or if a less than comprehensive neck dissection was performed [54]. However, while a recent large randomized study of adjuvant radiation therapy for metastatic melanoma showed that it is effective for reducing nodal relapse, this therapy failed to provide any survival benefit [55]. The role of radiation therapy in the adjuvant treatment of stage III melanoma remains uncertain.

The only systemic therapy with US Food and Drug Administration (FDA) approval for the treatment of resected stage III melanoma is interferon, which is indicated after completion of treatment of stage III disease. However, its efficacy is modest at best, and the toxicity associated with the usual year-long course of treatment with this agent can be significant [56].

The identification of several important melanoma-related gene mutations has ushered a new era of precision cancer treatment for stage IV melanoma [57]. BRAF, NRAS, and c-KIT have been the focus of implementation of targeted treatments for this disease. Approximately half of cutaneous melanomas possess the V600 mutation in the serine-threonine protein kinase BRAF. Clinical trials of the BRAF kinase inhibitor vemurafenib have demonstrated response rates of more than 50% for patients expressing the BRAF V600E mutation [58]. Vemurafenib significantly reduces the rate of disease progression and improves overall survival in stage IV melanoma. Ipilimumab, a monoclonal antibody that blocks the cytotoxic T-lymphocyte antigen CTLA-4, has been shown to improve the median overall survival of patients with stage IV melanoma [59].

The roles of these new targeted agents as primary or adjuvant treatments for stage III melanoma, specifically including metastatic disease involving the parotid gland, cannot be stated at this time. However, the poor prognosis of patients with regionally metastatic melanoma and the limited efficacy of surgery and currently available adjuvant therapies for preventing progression to stage IV disease highlight the need to develop better treatment protocols for this group of patients. Exploring the application of targeted agents that have been proven to be successful for stage IV melanoma and the development of other novel agents seems to be an important priority for future investigations if we are to achieve better success in the treatment of melanoma metastatic to the parotid and regional neck nodes.

References

1 Australian Cancer Network, Management of Non-Melanoma Skin Cancer Working Party: Non-Melanoma Skin Cancer: Guidelines for Treatment and Management in Australia. Canberra, National Health and Medical Research Council, 2002.
2 Buettner PG, Raasch BA: Incidence rates of skin cancer in Townsville, Australia. Int J Cancer 1998;78:587–593.
3 Australian Institute of Health and Welfare (AIHW): Australian Cancer Incidence and Mortality (ACIM) books: Non-Melanoma Skin Cancer, All Types. Canberra, AIHW, 2015. http://www.aihw.gov.au/acim-books (accessed May 2015).

4 Brantsch KD, Meisner C, Schönfisch B, et al: Analysis of risk factors determining prognosis of cutaneous squamous-cell carcinoma: a prospective study. Lancet Oncol 2008;9:713–720.

5 Clayman GL, Lee JJ, Holsinger FC, et al: Mortality risk from squamous cell skin cancer. J Clin Oncol 2005;23:759–765.

6 Ch'ng S, Tan ST, Davis P, et al: Skin cancers and the cell cycle; in Chen KL (ed): Progress in Cell Cycle Control Research. New York, Nova Science Publishers, 2008.

7 D'Souza J, Clark JR: Management of the neck in metastatic cutaneous squamous cell carcinoma of the head and neck. Curr Opin Otolaryngol Head Neck Surg 2011;19:99–105.

8 Veness MJ, Porceddu S, Palme CE, et al: Cutaneous head and neck squamous cell carcinoma metastatic to parotid and cervical lymph nodes. Head Neck 2007;29:621–631.

9 Gore SM, Shaw D, Martin RCW, et al: Prospective study of sentinel node biopsy for high-risk cutaneous squamous cell carcinoma of the head and neck. Head Neck DOI: 10.1002/hed.24120.

10 Ahmed MM, Moore BA, Schmalbach CE: Utility of head and neck cutaneous squamous cell carcinoma sentinel node biopsy: a systematic review. Otolaryngol Head Neck Surg 2014;150:180–187.

11 Schmitt AR, Brewer JD, Bordeaux JS, et al: Staging for cutaneous squamous cell carcinoma as a predictor of sentinel lymph node biopsy results: meta-analysis of American Joint Committee on Cancer criteria and a proposed alternative system. JAMA Dermatol 2014;150:19–24.

12 Martinez JC, Clark CO, Stasko T, et al: Defining the clinical course of metastatic skin cancer in organ transplant recipients. Arch Dermatol 2003;139:301–306.

13 Brunner M, Ng BC, Veness MJ, et al: Comparison of the AJCC N staging system in mucosal and cutaneous squamous head and neck cancer. Laryngoscope 2014;124:1598–1602.

14 TNM Classification of Malignant Tumours, ed 7. Union for International Cancer Control, 2010. www.uicc.org (accessed May 2015).

15 O'Brien CJ: The parotid gland as a metastatic basin for cutaneous cancer. Arch Otolaryngol Head Neck Surg 2005;131:551–555.

16 Andruchow JL, Veness MJ, Morgan GJ, et al: Implications for clinical staging of metastatic cutaneous squamous carcinoma of the head and neck based on a multicenter study of treatment outcomes. Cancer 2006;106:1078–1083.

17 Forest VI, Clark JR, Veness MJ, et al: N1S3: a revised staging system for head and neck squamous cell carcinoma with lymph node metastases. Cancer 2010;116:1298–1304.

18 Clark JR, Rumcheva P, Veness MJ: Analysis and comparison of the 7th edition American Joint Committee on Cancer (AJCC) nodal staging system for metastatic cutaneous squamous cell carcinoma of the head and neck. Ann Surg Oncol 2012;19:4252–4558.

19 Oddone N, Morgan GJ, Palme CE, et al: Metastatic cutaneous squamous cell carcinoma of the head and neck: the Immunosuppression, Treatment, Extranodal spread, and Margin status (ITEM) prognostic score to predict outcome and the need to improve survival. Cancer 2009;115:1883–1891.

20 Kelder W, Ebrahimi A, Forest V-I, et al: Cutaneous head and neck squamous cell carcinoma with regional metastases: the prognostic importance of soft tissue metastases and extranodal spread. Ann Surg Oncol 2012;19:274–279.

21 Vauterin TJ, Veness MJ, Morgan GJ, et al: Patterns of lymph node spread of cutaneous squamous cell carcinoma of the head and neck. Head Neck 2006;28:785–791.

22 Ch'ng S, Pinna A, Ioannou K, et al: Assessment of second tier lymph nodes in melanoma and implications for extent of elective neck dissection in metastatic cutaneous malignancy of the parotid. Head Neck 2013;35:205–208.

23 Ebrahimi A, Moncrieff MD, Clark JR, et al: Predicting the pattern of regional metastases from cutaneous squamous cell carcinoma of the head and neck based on location of the primary. Head Neck 2010;32:1288–1294.

24 Kirke DN, Porceddu S, Wallwork BD, et al: Pathologic occult neck disease in patients with metastatic cutaneous squamous cell carcinoma to the parotid. Otolaryngol Head Neck Surg 2011;144:549–551.

25 Jambursaria-Pahlajani A, Miller CJ, Quon H, et al: Surgical monotherapy versus surgery plus adjuvant radiotherapy in high-risk cutaneous squamous cell carcinoma: a systematic review of outcomes. Dermatol Surg 2009;35:574–585.

26 Veness MJ, Morgan GJ, Palme CE, et al: Surgery and adjuvant radiotherapy in patients with cutaneous head and neck squamous cell carcinoma metastatic to lymph nodes: combined treatment should be considered best practice. Laryngoscope 2005;115:870–875.

27 Ebrahimi A, Clark JR, Lorincz BB, et al: Metastatic head and neck cutaneous squamous cell carcinoma: defining a low-risk patient. Head Neck 2012;34:365–370.

28 Post-operative concurrent chemo-radiotherapy versus postoperative radiotherapy in high-risk cutaneous squamous cell carcinoma of the head and neck. (Post-Operative Skin Trial, POST study). Trans-Tasman Radiation Oncology Group, TROG 05.01.2009. www.trog.com.au/TROG-0501-POST (accessed August 2015).

29 American Cancer Society: Cancer Facts & Figures 2015. Atlanta, American Cancer Society, 2015.

30 Simard EP, Ward EM, Siegel R, et al: Cancers with increasing incidence trends in the United States: 1999 through 2008. CA Cancer J Clin 2012;62:118–128.

31 Harris TJ, Hinckley DM: Melanoma of the head and neck in Queensland. Head Neck Surg 1983;5:197–203.

32 Marzuka-Alcalá A, Gabree MJ, Tsao H: Melanoma susceptibility genes and risk assessment. Methods Mol Biol 2014;1102:381–393.

33 Balch CM, Soong SJ, Gershenwald JE, et al: Prognostic factors analysis of 17,600 melanoma patients: validation of the American Joint Committee on Cancer melanoma staging system. J Clin Oncol 2001;19:3622–3634.

34 Suton P, Lukšić I, Müller D, et al: Lymphatic drainage patterns of head and neck cutaneous melanoma: does primary melanoma site correlate with anatomic distribution of pathologically involved lymph nodes? Int J Oral Maxillofac Surg 2012;41:413–420.

35 Newlands C, Gurney B: Management of regional metastatic disease in head and neck cutaneous malignancy; cutaneous malignant melanoma. Br J Oral Maxillofac Surg 2014;52:301–307.

36 Testori A, Lazzaro G, Baldini F, et al: The role of ultrasound of sentinel nodes in the pre- and post-operative evaluation of stage I melanoma patients. Melanoma Res 2005;15:191–198.
37 Marone U, Catalano O, Caracò C, et al: Can high-resolution ultrasound avoid the sentinel lymph-node biopsy procedure in the staging process of patients with stage I-II cutaneous melanoma? Ultraschall Med 2012;33:E179–E185.
38 Hansson J, Ringborg U, Lagerlöf B, et al: Elective lymph node dissection in stage I cutaneous malignant melanoma of the head and neck; a report from the Swedish Melanoma Study Group. Melanoma Res 1994;4:407–411.
39 Fisher SR: Elective, therapeutic, and delayed lymph node dissection for malignant melanoma of the head and neck: analysis of 1,444 patients from 1970 to 1998. Laryngoscope 2002;112:99–110.
40 O'Brien CJ, Gianoutsos MP, Morgan MJ: Neck dissection for cutaneous malignant melanoma. World J Surg 1992;16:222–226.
41 Morton DL, Thompson JF, Cochran AJ, et al: Sentinel-node biopsy or nodal observation in melanoma. N Engl J Med 2006;355:1307–1317.
42 Schmalbach CE, Nussenbaum B, Rees RS, et al: Reliability of sentinel lymph node mapping with biopsy for head and neck cutaneous melanoma. Arch Otolaryngol Head Neck Surg 2003;129:61–65.
43 Parrett BM, Kashani-Sabet M, Singer MI, et al: Long-term prognosis and significance of the sentinel lymph node in head and neck melanoma. Otolaryngol Head Neck Surg 2012;147:699–706.

44 Balch CM, Gershenwald JE: Clinical value of the sentinel-node biopsy in primary cutaneous melanoma. N Engl J Med 2014;370:663–664.
45 Main BG, Coyle MJ, Godden A, et al: The metastatic potential of head and neck cutaneous malignant melanoma: is sentinel node biopsy useful? Br J Oral Maxillofac Surg 2014;52:340–343.
46 Loree TR, Tomljanovich PI, Cheney RT, et al: Intraparotid sentinel lymph node biopsy for head and neck melanoma. Laryngoscope 2006;116:1461–1464.
47 Picon AI, Coit DG, Shaha AR, et al: Sentinel lymph node biopsy for cutaneous head and neck melanoma: mapping the parotid gland. Ann Surg Oncol DOI: 10.1245/ASO.2006.03.051.
48 Samra S, Sawh-Martinez R, Tom L, et al: A targeted approach to sentinel lymph node biopsies in the parotid region for head and neck melanomas. Ann Plast Surg 2012;69:415–417.
49 Vermeeren L, Valdés Olmos RA, Klop WM, et al: SPECT/CT for sentinel lymph node mapping in head and neck melanoma. Head Neck 2011;33:1–6.
50 van Akkooi AC, Voit CA, Verhoef C, et al: New developments in sentinel node staging in melanoma: controversies and alternatives. Curr Opin Oncol 2010;22:169–177.
51 Niebling MG, Bastiaannet E, Hoekstra OS, et al: Outcome of clinical stage III melanoma patients with FDG-PET and whole-body CT added to the diagnostic workup. Ann Surg Oncol 2013;20:3098–3105.

52 Thom JJ, Moore EJ, Price DL, et al: The role of total parotidectomy for metastatic cutaneous squamous cell carcinoma and malignant melanoma. JAMA Otolaryngol Head Neck Surg 2014;140:548–554.
53 Ch'ng S, Pinna A, Ioannou K, et al: Assessment of second tier lymph nodes in melanoma and implications for extent of elective neck dissection in metastatic cutaneous malignancy of the parotid. Head Neck 2013;35:205–208.
54 Moncrieff MD, Martin R, O'Brien CJ, et al: Adjuvant postoperative radiotherapy to the cervical lymph nodes in cutaneous melanoma: is there any benefit for high-risk patients? Ann Surg Oncol 2008;15:3022–3027.
55 Burmeister BH, Henderson MA, Ainslie J, et al: Adjuvant radiotherapy versus observation alone for patients at risk of lymph-node field relapse after therapeutic lymphadenectomy for melanoma: a randomised trial. Lancet Oncol 2012;13:589–597.
56 Mocellin S, Lens MB, Pasquali S, et al: Interferon alpha for the adjuvant treatment of cutaneous melanoma. Cochrane Database Syst Rev 2013;6:CD008955.
57 Walkington LA, Lorigan P, Danson SJ: Advances in the treatment of late stage melanoma. BMJ 2013;346:f1265.
58 Chapman PB, Hauschild A, Robert C, et al: Improved survival with vemurafenib in melanoma with BRAF V600E mutation. N Engl J Med 2011;364:2507–2516.
59 Graziani G, Tentori L, Navarra P: Ipilimumab: a novel immunostimulatory monoclonal antibody for the treatment of cancer. Pharmacol Res 2012;65:9–22.

Prof. Jonathan Clark
Chris O'Brien Lifehouse
Missenden Road, PO Box M33
Camperdown, Sydney, NSW 2050 (Australia)
E-Mail Jonathan.Clark@lh.org.au

Surgery for Malignant Submandibular Gland Neoplasms

Natalie L. Silver[a] · Steven B. Chinn[a] · Patrick J. Bradley[b] · Randal S. Weber[a]

[a]UT MD Anderson Cancer Center, Department of Head and Neck Surgery, Houston, Tex., USA; [b]School of Medicine, The University of Nottingham, Nottingham University Hospitals, Queens Medical Centre Campus, Nottingham, UK

Abstract

For many decades, surgery has been the primary treatment for malignant submandibular gland neoplasms. Nonetheless, due to the heterogeneity and rarity of submandibular gland malignant tumors and the high frequency of chronic benign processes in this region, management can be complex. Preoperative investigations, such as fine-needle aspiration and imaging, are critical to achieve the correct diagnosis so that appropriate surgery can be planned. In general, for malignant submandibular gland neoplasms, the minimal treatment necessary is excision of the submandibular gland with level I lymph node dissection. Salivary gland cancer in the submandibular gland is generally more aggressive than the same histologic type in the parotid gland. Neck dissection may be required and primarily depends on the stage and histological grade. Adjuvant therapy most frequently consists of radiation and can improve overall survival. Some factors that influence prognosis after surgical treatment include the histologic grade, stage at presentation, and positive surgical margins.

© 2016 S. Karger AG, Basel

Introduction

Salivary gland tumors may originate in either the major or minor salivary glands, and the proportion of benign to malignant salivary gland neoplasms is dependent on the location. A mass in the submandibular gland or a minor salivary gland is more likely to be malignant. In a review of more than two thousand salivary gland tumor cases, 73% of the tumors were found in the parotid gland, with only 15% found to be malignant, while 11% were found in the submandibular gland, with 37% found to be malignant [1].

It is imperative for physicians to be able to distinguish a chronic benign process, such as sialadenitis, from a submandibular gland neoplasm and then to further determine whether the neoplasm is benign or malignant. This is done through a careful history and physical examination, as well as by preoperative imaging and fine-needle aspiration (FNA).

Management of submandibular gland malignancy can be challenging due the relative rarity of the disease and the diversity of its behavior due to the existence of a variety of histologic subtypes and grades. Adenoid cystic carcinoma is the most common subtype in the submandibular gland, followed by mucoepidermoid carcinoma and then adenocarcinoma. Table 1 displays the histologic spectrum of malignancy for tumors of the submandibular gland observed at our institution [2, 3].

Clinical Presentation

Most often, a tumor of the submandibular gland presents as a mass or swelling. For patients with chronic sialadenitis, the typical symptoms include intermittent swelling of the submandibular gland associated with eating or drinking. The gland may become firm and painful. Benign and malignant tumors of the submandibular gland can both present as a painless mass. However, a neoplasm accompanied by pain is suggestive of malignancy. In up to 20% of patients with malignant tumors, pain may be constant and progressive, while benign neoplasms rarely present with pain [4]. Gradual enlargement is also more common in patients with submandibular neoplasia, and progressive enlargement over a short period of time suggests malignancy (fig. 1a).

Physical examination includes inspection and palpation of the neck as well as the oral cavity. Bimanual palpation should be performed to determine whether the gland is fixated to an adjacent structure, such as the mandible or skin. It is important to assess sensation of the tongue (indicating lingual nerve involvement), tongue fasciculations or weakness (indicating hypoglossal nerve involvement) and lip weakness (indicating marginal mandibular nerve involvement), all of which indicate perineural spread of the tumor. The presence or absence of trismus should be assessed, which, if present, indicates invasion of the medial pterygoid muscle.

Careful examination of the neck is critical for identifying lymphadenopathy, particularly in level I, because adenopathy with an enlarged submandibular gland is highly suspicious of malignancy. At initial presentation, lymph node metastasis from a submandibular gland malignancy is present in between 8 and 35% of patients [4–6].

Table 1. Histology of malignant submandibular gland tumors

Histologic type	Patients, n (%)
Adenoid cystic	57 (66)
Mucoepidermoid carcinoma	15 (17)
Adenocarcinoma	5 (6)
Undifferentiated	3 (3)
Acinic cell	2 (2)
Lymphoepithelioma	2 (2)
Squamous cell	1 (1)
Carcinoma ex-pleomorphic adenoma	1 (1)
Other	1 (1)
Total	87

Imaging Studies

Imaging for the assessment of submandibular gland lesions augments the physical examination and aids in the following: determining whether a lesion is intrinsic or extrinsic to the gland, evaluating the extent of the lesion with respect to local invasion, establishing perineural involvement, and determining whether there is metastatic disease. All of these factors are important to delineate prior to surgical management so that the appropriate procedures are discussed and planned for [7].

Although imaging lacks the specificity to distinguish benign from malignant tumors, a CT scan with contrast can provide valuable information regarding mandibular bone invasion, the local extent of the tumor, and the presence or absence of pathologic lymphadenopathy (fig. 1b). MRI can provide superior soft tissue detail com-

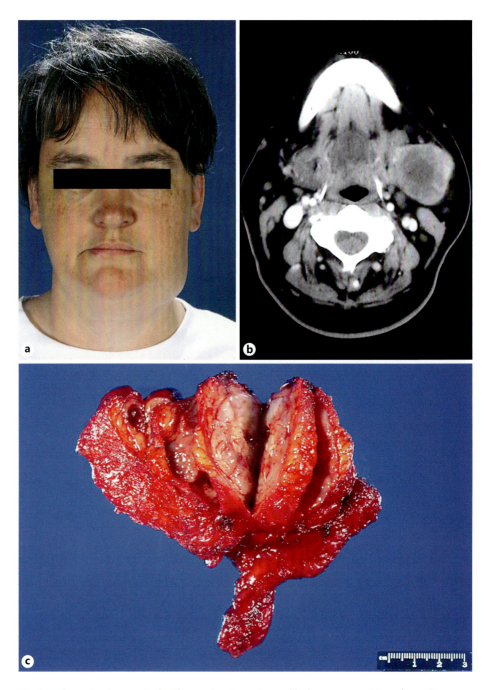

Fig. 1. a The patient presented with an enlarging submandibular mass suspicious for malignancy. **b** CT scan with contrast revealed a heterogeneously enhancing mass of the submandibular gland that did not appear to be locally invasive. **c** The tumor was completely excised with level I lymph node dissection, and final pathology revealed carcinosarcoma of the submandibular gland with no positive lymph nodes or perineural invasion.

pared with CT scans and can help to assess perineural spread.

Positron emission tomography (PET)-CT in salivary gland disease can help to rule out distant metastasis if the primary cancer has enhanced fluorodeoxyglucose (FDG) uptake. In FDG-avid cancers, PET-CT may be useful for the initial staging, histologic grading and monitoring for recurrence [8, 9]. However, an inflamed or infected submandibular gland may also uptake FDG and show enhancement on PET-CT.

Fine-Needle Aspiration Biopsy

Preoperative FNA, performed as a diagnostic test, is a vital part of the clinical management algorithm for submandibular gland diseases. For inflammatory causes of salivary gland enlargement, nonsurgical management can often be used, or a simple submandibular gland excision may be planned without level I neck dissection. Therefore, when properly combined with clinical-radiologic findings, FNA results (which typically demonstrate the presence of acute and chronic inflammatory cells) can aid in surgical planning and can allow cultures to be obtained from suspected infectious masses. Regarding malignancy, if the histopathological type can be determined in advance, this information may be used for preoperative counseling regarding the extent of surgery, primarily regarding the necessity of neck dissection. However, differentiating between benign and malignant salivary gland tumors can be more difficult. Nonetheless, when performed by experienced practitioners, FNA provides accurate results in over 90–95% of patients [10]. The use of FNA in the work-up of salivary tumors is also cost effective. Layfield et al. [11] have demonstrated that the use of routine FNA in the work-up of salivary gland lesions saves up to USD 70,000 per 100 patients and reduces operative interventions by 65% for submandibular masses and by 35% for parotid masses.

Ultrasound-guided biopsy can be useful for the diagnosis of suspicious heterogeneous glands. Additionally, core needle biopsy is more sensitive and specific than FNA in diagnosing malignant lesions and can therefore be used as an additional diagnostic tool for uncertain lesions, especially in patients who may require extensive surgery that must be discussed in advance [12].

Histopathology

Salivary gland malignancies are extraordinarily heterogeneous and complex in histology, resulting in variable clinical behaviors and therefore clinical management. The 2005 World Health Organization (WHO) classifications described 24 different salivary gland phenotypes, and the same TNM staging classification is used for all histologic types of salivary gland cancer arising in the major salivary glands [13].

Adenoid Cystic Carcinoma

Adenoid cystic carcinoma is the most common type of malignant tumor of the submandibular gland. It is characterized by a locally infiltrative growth pattern with perineural invasion and a high rate of local recurrence and delayed distant metastasis [14, 15]. Perineural invasion can be seen in up to 75% of cases and can affect the extent of surgery, requiring sacrifice of nerves if a tumor is involved, most often the lingual and hypoglossal nerves [16]. Perineural spread may also involve the marginal mandibular nerve or the cervical branch of the facial nerve. Adenoid cystic carcinoma is more aggressive in the submandibular gland than in the parotid gland. A solid tumor subtype, lymphovascular invasion, and positive margins are also correlated with a poorer prognosis [17, 18]. Despite the ability to achieve good initial local control, adenoid cystic carcinoma is the most common type of salivary gland carcinoma associated with distant metastasis, which can develop many years after initial diagnosis [19]. In addition, although the majority of patients with clinically early-stage adenoid cystic carcinoma of the salivary glands have a favorable prognosis, a significant percentage (20%) will develop distant metastasis and therefore need to be monitored carefully over a longer period of time [20].

Mucoepidermoid Carcinoma

Mucoepidermoid carcinoma is the second most common malignant tumor of the submandibular gland and can have a low-, intermediate-, or high-grade histology. High-grade lesions, an advanced stage, perineural invasion, positive margins and a submandibular gland location are all associated with a worse prognosis. There is a higher proportion of intermediate- or high-grade lesions in patients presenting with a submandibular gland primary lesion, and up to 50% of mucoepidermoid carcinomas located in this region will have cervical metastasis compared to 28% for the parotid gland [21]. High-grade tumors are more aggressive, can invade locally, and are also more likely to have nodal metastasis [22]. However, even low-grade mucoepidermoid carcinoma of the submandibular gland can recur and metastasize more frequently than carcinoma of the parotid or minor salivary glands, necessitating aggressive resection of any mucoepidermoid carcinoma primary lesion in this location [21, 23].

Adenocarcinoma

Adenocarcinoma of the salivary glands encompasses a wide array of histopathological entities, such as salivary duct carcinoma, adenocarcinoma (not otherwise specified), polymorphous low-grade adenocarcinoma and basal cell carcinoma. It is generally differentiated into low- and high-grade histologies with salivary duct carcinoma, and about half of adenocarcinoma (not otherwise specified) cases are high-grade entities. Overall survival is low at 43% over 5 years and is associated with several clinic-pathological factors, such as a fixed mass or rapid tumor growth, a diagnosis of adenocarcinoma (not otherwise specified), and positive surgical margins [24].

Treatment

Surgery is the primary treatment for patients with resectable salivary gland cancer. The minimal procedure performed should be complete excision of the gland and levels Ia and Ib lymph node dissection, with a careful attempt to spare uninvolved nerves. Generally, this is an acceptable treatment for tumors that are low grade and early stage (i.e. no clinical or radiographic neck disease). High-grade, advanced-stage tumors require more extensive surgery involving ipsilateral selective neck dissection in addition to excision of the submandibular gland. When a tumor is locally advanced, resection of adjacent structures, such as the mandible, involved nerves, or skin, may be necessary (fig. 2). In these cases, reconstruction may be required with local, regional or free microvascular flaps [4, 7].

Several clinic-pathological factors are considered when recommending adjuvant treatment. The indications are as follows: positive surgical margins, high-grade histology, locally advanced disease (perineural/bone invasion), and advanced stage. Postoperative treatment generally consists of ipsilateral neck irradiation and can significantly increase local control compared with surgery alone [25]. Garden et al. [17] treated patients with adenoid cystic carcinoma of the submandibular gland and suspected microscopic residual disease with postoperative radiotherapy and achieved a 10-year survival rate of ~60%. The role of adjuvant chemoradiotherapy in the treatment of resected high-risk salivary gland lesions is currently under investigation (Radiation Therapy Oncology Group, Protocol #1008).

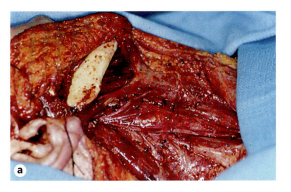 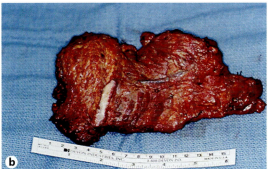

Fig. 2. a The patient presented with recurrent high-grade mucoepidermoid carcinoma of the submandibular gland requiring wide local excision of the gland with removal of tissues from the floor of mouth and a previous scar, and radical neck dissection was also performed. **b** The entire specimen was removed en bloc.

At our institution, of 86 patients with submandibular gland malignancy treated initially with surgery, 45% experienced recurrences, of which half were loco-regional recurrences and the other half occurred at distant sites. The 2- and 5-year survival rates were 82 and 69%, respectively [4]. Loco-regional control was enhanced by adjuvant radiation.

Surgical Technique for Excision of Malignant Submandibular Gland Tumors

The patient is placed supine, and general endotracheal anesthesia is induced. A shoulder roll is placed to extend the neck. The neck and face are prepped and draped with adequate exposure of the submental area and corner of the mouth. An incision is marked in a natural skin crease at least two fingerbreadths below the edge of the mandible to protect the marginal mandibular nerve (fig. 3a). The incision should extend from the anterior boarder of the sternocleidomastoid muscle to the submental region. The skin and platysma are incised, the superior flap is elevated to the inferior border of the mandible, and the inferior flap is elevated below the submandibular gland (fig. 3b). If the platysma is infiltrated with tumor, flap elevation should occur in a supraplatysmal plane, with excision of the involved muscle. The marginal mandibular nerve is identified, traced over the mandible, and elevated superiorly. The facial artery and vein are clamped and ligated inferior to the marginal mandibular nerve. The peri-facial lymph nodes, proximal facial vessels and lateral aspects of the submandibular gland are reflected inferiorly and anteriorly. The mylohyoid muscle is skeletonized to its inferior edge near the digastric tendon, including all of the fatty areolar tissue. If a tumor is fixed to the mandible without evidence of invasion, then a rim of the mandible may be excised. Frank invasion necessitates mandibulectomy.

Next, the submental region is dissected. The medial edge of the contralateral anterior belly of the digastric muscle is first skeletonized. The submental lymph nodes are reflected inferiorly off of the underlying mylohyoid muscle. If any muscle is infiltrated with tumor, then it is resected with the gland. The specimen is reflected toward the ipsilateral anterior belly of the digastric muscle and also skeletonized. The level Ia specimen should be continuous with the level Ib contents. The mylohyoid muscle is then retracted, exposing the lingual nerve and deeper portion of the submandibular gland. When perineural in-

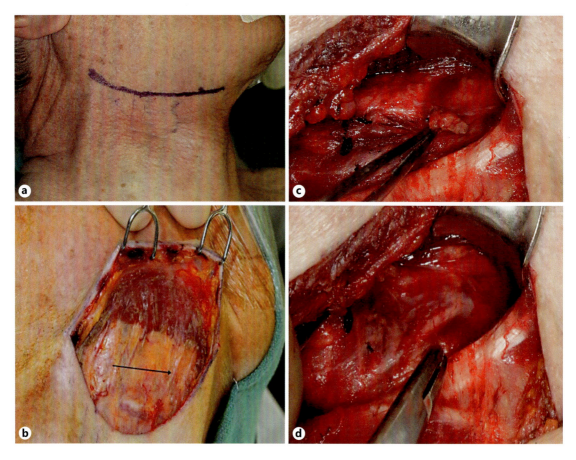

Fig. 3. a An incision is marked at least two fingerbreadths below the edge of the mandible. **b** Subplatysmal flaps are elevated. The arrow points to the marginal mandibular nerve. **c** The lingual nerve is visible. The submandibular ganglion and duct have been ligated. **d** The hypoglossal nerve here is visible deep into the ligated submandibular duct. Rights to reprint these pictures have been obtained from Wolters Kluwer Health (License #3677830789655).

vasion is suspected, frozen section samples may be sent intra-operatively, and if nerves are positive for tumor, then they are resected with the gland. The nerves should be followed in a retrograde fashion until clear margins can be obtained. Of note, adenoid cystic carcinoma often has skip metastasis, making frozen section analysis less reliable when attempting to obtain clear margins.

The submandibular ganglion is next clamped, cut, and ligated (fig. 3c). The submandibular duct is identified and followed. The duct is ligated, and the surrounding portion of the submandibular gland is reflected inferiorly with the attached specimen. Prior to ligating the duct, the hypoglossal nerve should be identified deep into the duct and preserved if it is uninvolved by tumor (fig. 3d). The specimen is reflected posteriorly, where it remains tethered by the proximal portion of the facial artery. The artery is clamped, cut, and ligated. The wound is irrigated, and a suction drain is placed. The wound is closed in several layers [3].

Surgical Management of the Neck

A clinically positive neck should be treated by ipsilateral neck dissection, including all grossly involved nodes, while attempting to spare vital structures. The most common involved nodes in carcinoma of the submandibular gland are located in levels I–III, but the skipping of levels can occur; therefore, comprehensive neck dissection is recommended in the setting of clinically positive disease [22].

The management of the N0 neck is controversial. Options include observation, elective neck dissection or radiation. As mentioned previously, all malignant tumors in the submandibular gland mandate levels Ia and Ib dissection. However, elective neck dissection is typically reserved for tumors with a high propensity for occult metastasis [discussed further in the chapter by Medina et al., this vol., pp. 132–140].

Conclusions

Submandibular gland malignancies are rare and present both a treatment and diagnostic challenge due to their histologic heterogeneity and aggressiveness. Complete submandibular gland excision with regional dissection of the submental lymph nodes is the primary treatment for malignant submandibular gland tumors. Adjuvant radiotherapy can decrease the local recurrence rate and improve survival in patients with adverse features while adjuvant chemotherapy is being investigated. It is imperative to follow patients long after initial treatment is provided due to the high likelihood of distant metastasis.

References

1 Eveson JW, Cawson RA: Salivary gland tumours. A review of 2410 cases with particular reference to histological types, site, age and sex distribution. J Pathol 1985;146:51–58.
2 Kaszuba SM, Zafereo ME, Rosenthal DI, et al: Effect of initial treatment on disease outcome for patients with submandibular gland carcinoma. Arch Otolaryngol Head Neck Surg 2007;133:546–550.
3 Weber R: Excision of the submandibular gland; in Ferris RL (ed): Master Techniques in Otolaryngology Head and Neck Surgery. Philadelphia, Lippincot Williams and Wilkins, 2013, pp 261–269.
4 Weber RS, Byers RM, Petit B, et al: Submandibular gland tumors. Adverse histologic factors and therapeutic implications. Arch Otolaryngol Head Neck Surg 1990;116:1055–1060.
5 Spiro RH: Salivary neoplasms: overview of a 35-year experience with 2,807 patients. Head Neck Surg 1986;8:177–184.
6 Bhattacharyya N: Survival and prognosis for cancer of the submandibular gland. J Oral Maxillofac Surg 2004;62:427–430.
7 Dias F: Management of tumors of the submandibular and sublingual glands; in Myers E (ed): Salivary Gland Disorders. New York, Springer, 2007, pp 339–374.
8 Razfar A, Heron DE, Branstetter BF 4th, et al: Positron emission tomography-computed tomography adds to the management of salivary gland malignancies. Laryngoscope 2010;120:734–738.
9 Roh JL, Ryu CH, Choi SH, et al: Clinical utility of 18F-FDG PET for patients with salivary gland malignancies. J Nucl Med 2007;48:240–246.
10 Stewart CJ, MacKenzie K, McGarry GW, et al: Fine-needle aspiration cytology of salivary gland: a review of 341 cases. Diagn Cytopathol 2000;22:139–146.
11 Layfield LJ, Gopez E, Hirschowitz S: Cost efficiency analysis for fine-needle aspiration in the workup of parotid and submandibular gland nodules. Diagn Cytopathol 2006;34:734–738.
12 Novoa E, Gurtler N, Arnoux A, et al: Diagnostic value of core-needle biopsy and fine-needle aspiration in salivary gland lesions. Head Neck DOI: 10.1002/hed.23999.
13 Eveson JW: WHO histological classification of tumours of the salivary glands; in Barnes L (ed): World Health Organization Classification of Tumours: Pathology and Genetics of Head and Neck Tumours. Lyon, IARC Press, 2005, p 210.
14 Sung MW, Kim KH, Kim JW, et al: Clinicopathologic predictors and impact of distant metastasis from adenoid cystic carcinoma of the head and neck. Arch Otolaryngol Head Neck Surg 2003;129:1193–1197.
15 Spiro RH: Distant metastasis in adenoid cystic carcinoma of salivary origin. Am J Surg 1997;174:495–498.
16 Spiro RH, Huvos AG, Strong EW: Adenoid cystic carcinoma of salivary origin. A clinicopathologic study of 242 cases. Am J Surg 1974;128:512–520.

17 Garden AS, Weber RS, Morrison WH, et al: The influence of positive margins and nerve invasion in adenoid cystic carcinoma of the head and neck treated with surgery and radiation. Int J Radiat Oncol Biol Phys 1995;32:619–626.

18 Cohen AN, Damrose EJ, Huang RY, et al: Adenoid cystic carcinoma of the submandibular gland: a 35-year review. Otolaryngol Head Neck Surg 2004;131: 994–1000.

19 Bradley PJ: Distant metastases from salivary glands cancer. ORL J Otorhinolaryngol Relat Spec 2001;63:233–242.

20 Bhayani MK, Yener M, El-Naggar A, et al: Prognosis and risk factors for early-stage adenoid cystic carcinoma of the major salivary glands. Cancer 2012;118: 2872–2878.

21 Spiro RH, Huvos AG, Berk R, et al: Mucoepidermoid carcinoma of salivary gland origin. A clinicopathologic study of 367 cases. Am J Surg 1978;136:461–468.

22 McHugh CH, Roberts DB, El-Naggar AK, et al: Prognostic factors in mucoepidermoid carcinoma of the salivary glands. Cancer 2012;118:3928–3936.

23 Luna MA: Salivary mucoepidermoid carcinoma: revisited. Adv Anat Pathol 2006;13:293–307.

24 Huang AT, Tang C, Bell D, et al: Prognostic factors in adenocarcinoma of the salivary glands. Oral Oncol 2015;51: 610–615.

25 Terhaard CHJ, Lubsen H, Rasch CRN, et al: The role of radiotherapy in the treatment of malignant salivary gland tumors. Int J Radiat Oncol 2005;61:103–111.

Natalie L. Silver, MD, MS
Department of Head and Neck Surgery Unit 1445, MD Anderson Cancer Center
1515 Holcombe Blvd
Houston, TX 77030 (USA)
E-Mail nsilver@mdanderson.org

Surgery for Malignant Sublingual and Minor Salivary Gland Neoplasms

Patrick J. Bradley[a] · Robert L. Ferris[b]

[a]School of Medicine, The University of Nottingham, Nottingham University Hospitals, Queens Medical Centre Campus, Nottingham, UK; [b]Chief Division of Head and Neck Surgery, Department of Otolaryngology, University of Pittsburgh School of Medicine, Pittsburgh, Pa., USA

Abstract

Malignant sublingual gland neoplasms are rare, early-stage neoplasms presenting as painless non-ulcerated masses in the antero-lateral floor of the mouth. The majority of patients present with advanced disease, with symptoms of pain or anaesthesia of the tongue. Malignant minor salivary gland neoplasms are more common, the majority (>80%) of which present in the oral cavity, most frequently in the palatal area, as painless masses or as obstructive symptoms in the head and neck region. The most frequent pathologies are adenoid cystic carcinoma and mucoepidermoid carcinoma (>85%), with the majority presenting at an advanced stage (III/IV). Wide tumour-free surgical margin excision is the treatment of choice, followed by radiotherapy, after discussion of the multidisciplinary head and neck cancer tumour board. Improvements in survival and quality of life have been achieved since the introduction of endoscopic and robotic surgeries for many minor salivary gland malignancies.

© 2016 S. Karger AG, Basel

Malignant Sublingual Gland Neoplasms

Anatomy of Sublingual Glands

The paired sublingual glands are the smallest of the major salivary glands (2–4 g). They have been described as almond or tadpole-shaped masses that are primarily composed of mucinous glands. The head of the tadpole, which is oval in shape, lies in the anterior floor of the mouth, and the tail, which is wedge shaped, runs posterior in the salivary gutter on the mylohyoid muscle. The sublingual gland is not a single organ but comprises a large segment (the major sublingual gland) and a group of 8–30 small, independent, aggregated glands (the minor sublingual glands). Thus, the excretory ducts may range in number from eight to twenty. Posteriorly, the sublingual gland contacts the deep aspect of the submandibular gland. The sublingual fossa of the mandible is located laterally, with the genioglossus muscle located medially. Other adjacent structures include the lingual nerve and vessels (superficial) and the hypoglossal nerve (deep), with the submandibular duct lying medially. The sublingual gland does not have a true fascial capsule layer.

Pathology

Adenoid cystic carcinoma is the most common (72%) malignant tumour of the sublingual gland, followed by mucoepidermoid carcinoma (MEC) (13%). Other malignant salivary gland pathologies have been reported [1–4]. Most patients range in age between 40 and 60 years, without gender predilection.

Evaluation and Staging

When presented with a submucosal mass in the floor of the mouth, imaging by ultrasound, computed tomography (CT) or magnetic resonance imaging (MRI) will allow for characterization of the consistency, extent and association with related structures, as well as the presence or absence of cervical nodal involvement. Biopsy for histopathological examination must be obtained by fine-needle aspiration cytology or fine-needle core biopsy before proceeding with or planning any treatment [discussed further in the chapter by Howlett and Triantafyllou, this vol., pp. 39–45]. Malignant epithelial sublingual gland neoplasms are staged in a similar manner to parotid and submandibular gland neoplasms [discussed further in the chapter by Bradley, this vol., pp. 1–8]. An advanced stage of disease at presentation is observed in >50% of patients, according to the few reported case series [1–4].

Management and Prognosis

The initial treatment is surgery, which is performed to encompass the extent of the primary tumour. Most series [2–4] recommend that tumours that are <2 cm in size and mobile can be resected intraorally but also recommend that the ipsilateral submandibular gland should be excised, as the ductal system is likely to be compromised by the disease or surgery. Tumours >2 cm in size should receive en bloc resection, possibly using a 'pull-through' procedure or a mandibulotomy approach. Repair of the surgical defect may be achieved by primary closure, but in wider resections, which may include bone and/or local soft tissue, the use of a local or myo-cutaneous flap or a free muscle-bone flap will enhance tissue repair and preserve tongue mobility. Marginal mandibulotomy is recommended when the periosteum alone is invaded, and segmental mandibulectomy may occasionally be necessary [3, 4]. Resection of the lingual nerve is often recommended, with frozen section analysis to ensure that the proximal end of the nerve is free of disease [3, 4].

Performing elective neck dissection (END) (levels I–IIa) may facilitate such a surgical procedure in the N0 situation. In N+ disease, therapeutic neck dissection (TND), such as modified neck dissection, should be performed. Adjuvant postoperative irradiation is indicated for patients with advanced-stage disease (stages III/IV), high-grade tumours, perineural invasion, positive surgical margins, >2 N+ lymph nodes and/or pathological evidence of extracapsular nodal extension.

The 5-year overall and disease free survival rates are approximately 75% and 85%, respectively [1], but OS continues to drop after 5 years, resulting in considerably lower 10- and 20-year survival rates [3]. Local recurrence, when it occurs (15–30%), has resulted in death in almost all cases that have been reported [2, 4]. Distant metastasis occurs in some patients, especially in those with adenoid cystic carcinoma.

Malignant Minor Salivary Gland Neoplasms

Introduction

Much has been published on malignant minor salivary gland neoplasms (MMSGNs), but it is difficult to summarize the literature due to the extreme heterogeneity of the patient cohorts studied and the time periods in which the data have been gathered, which are usually many years in duration, during which time pathology classifications have been revised and new tumour types have been added, and treatment modalities have also changed. In addition, in the past, major and minor salivary gland neoplasms were combined and the morphologic subtypes were mixed. All of these factors make it difficult for meaningful conclusions to be made [5].

Many of the 24 MMSGNs show significant morphological diversity, resulting in a number of overlapping features, which makes differentiation among tumour types difficult [6]. Particular attention should therefore be given to examination of an adequate tissue specimen rather than small forceps biopsy and/or punch biopsy specimens, and those from lesions at palatal, nasal and pharyngeal sites should be especially discouraged. The inclusion of the tumour's periphery is required in any biopsy to determine whether infiltrative growth is present or absent, indicating invasion into non-neoplastic seromucinous glands, soft tissues and/or bone [7].

Accurate estimation of the three-dimensional extent of anatomic disease is necessary for optimum surgical planning, and modern radiological imaging techniques, such as CT and/or MRI, are mandatory. T-max (the time to maximum contrast enhancement) of contrast-enhanced MRI is helpful for differentiating a benign lesion from MMSGN, but the results of further studies are being awaited [8]. The use of positron emission tomography, with or without CT, is becoming established in the staging of MMSGN at primary, nodal and metastatic sites and has been reported to be more specific than conventional imaging modalities [9].

All MMSGNs are staged using the TNM classification system by the defined anatomical site, similar to the staging of head and neck squamous cell carcinoma. This information in combination with other clinical, histopathological, patient and tumour features are used for treatment planning. The TNM components and stage groupings have repeatedly been found to be the strongest prognostic predictors [10].

Treatment

The treatment of choice for MMSGN is wide local excision with clear tumour margins. Resectability is determined pre-operatively based on an extensive imaging workup, the anatomic site, the tumour histology, and the local available expertise for performing head and neck surgery. These factors determine the extent of resection required and the functional implications of such a resection. In all series, the resection margin status has been shown to be an important prognostic factor that is strongly correlated with both the anatomic extent and histologic type [10].

Once a tumour has been resected and pathologically staged and graded, post-operative radiotherapy to the primary site is recommended for most patients. Only in early-stage disease with clear margins (stages I/II) and without adverse prognostic factors, such as lymphovascular or perineural invasion (a set of conditions usually restricted to low-grade variants), can post-operative radiotherapy be omitted safely [10]. Adverse prognostic factors, such as positive or close margins, high-grade malignancy, perineural growth (particularly a major named nerve) invasion, bone and muscle invasion, a paranasal sinus location, and high T and N classifications, are all indications for post-operative radiotherapy [11]. The indications for the inclusion of chemotherapy with planned radiotherapy post-operatively remain under investigation.

Currently, surgical treatment of the neck nodes is indicated only when there is clinical or radiologic evidence of regional metastasis, for which TND is performed, or when the risk of subclinical disease in a clinically negative neck exceeds 15–20%, for which END is performed. Clinical N+ disease presents in about 15% of patients [12]. Elective neck irradiation may be used as an alternative treatment for the negative neck on imaging when the primary site is receiving post-operative radiotherapy [13].

Site-Specific Surgical Treatment Considerations

Oral Cavity. The most common sites for tumours of MSGN origin are the palate, buccal mucosa, retromolar trigone, and upper lip, accounting for >75% of cases. The palate is the most common site of all MSGNs (55%), with >60% malignancy. The histopathology of the majority (>90%) of oral cavity MMSGNs is MEC or acinic cell carcinoma (ACC).

A review of a large series of 103 patients with MMSGN of the hard palate [14] showed involvement by ACC in 48 patients (47%), MEC in 37 (35.92%), carcinoma ex-pleomorphic adenoma in 15 (15%), and adenocarcinoma in 3 (3%). The outcomes at 5- and 10-years included overall survival (OS) rates of 77.9 and 65.7%, recurrence-free survival rates of 64.4 and 53.2%, and disease-specific survival rates of 77.9 and 67.7%, respectively. There were no significant differences in OS, recurrence-free survival or disease-specific survival between patients who underwent surgery alone and those who underwent surgery plus post-operative radiotherapy. Surgical excision with adequate margins was essential for a favourable outcome.

A review of the Surveillance Epidemiology and End Results (SEER) data collected between 1988 and 2005 from 639 patients with MMSGN of the oral cavity revealed a predominance of stages T1 and T4 (42.6 vs. 35.2%) and N0 disease in 93.4% [15]. The overall mean survival time was 157.9 months and was similar across histologic subtypes, with a poorer prognosis for high-grade ACC tumours. Another review of 68 patients with MMSGN reported that treatment with wide excision was the preferred procedure for the primary site [16]. The resection margin was recorded intra-operatively; R0 resection (complete gross and microscopic resection) was achieved in 74.7% of the patients, and R1 resection (gross total resection, but seemingly positive microscopic margins) was achieved in 25.3% of the patients, and no patients were recorded as having any remaining gross disease. Cut margins were negative (>5 mm away and free) in 70.5% of the patients, close (<5 mm away and free) in 8.5% and involved in 21%. Post-operative radiotherapy was completed for 40 patients (38.5%). Of the 87 patients with MMSGN, 13 (14.9%) had neck nodes on presentation and were treated with TND. Of these 87 patients, loco-regional recurrence was seen in 16 (18.4%), and it was more common in ACC (12/47; 25.5%) compared to non-ACC tumours (4/40; 10%). The estimated 5- and 10-year OS and disease-free survival (DFS) rates were 87.3 and 75.2% and 77.2 and 65.8%, respectively. When the DFS rates of the patients with ACC and non-ACC tumours were compared, the estimated 5- and 10-year rates were 66.1 and 44.1% for ACC and 89.4 and 81.3% for non-ACC. Similarly, the 5- and 10-year OS rates were better for the patients with non-ACC tumours (94.4 and 85.9%) compared with those with ACC tumours (80.8 and 60.6%) (log-rank test, p = 0.04), in agreement with the results of previous studies.

Oropharynx. The most frequent histologic types are ACC and MEC (>60%), followed by adenocarcinoma (not otherwise specified) [17]. Surgical resection is the mainstay of treatment; however, given the propensity for submucosal growth and relatively inaccessible location, as evidenced by the frequency of positive margins, patient outcomes have been poor. The need for adjuvant radiation depends on several factors, including the tumour size, histologic subtype and grade, margins, and nodal status. A series of 67 patients with MMSGN treated with conventional open surgery at the Memorial Sloan Kettering Cancer Centre [17] from 1985 to 2005 has reported on the factors predicting patient outcome. The most frequent site affected was the base of the tongue 41 patients (61%), the soft palate in 20 (30%) and the tonsils in 6 (9%). Fifty-three patients (79%) presented with T1/2 tumours, and 20 (30%) had cervical nodal metastasis at presentation. The patients with early- and late-stage nodal disease presented with MEC (9/21 and 3/5, respectively), ACC (1/13 and 0/3, respectively), and adenocarcinoma (not otherwise specified) (3/6 and 2/3, respectively). Of the 61 patients (91%) who underwent primary surgery, 20 (33%) received transoral surgery, 4 (7%) underwent transcervical surgery, and 37 (61%) received transmandibular surgery. The surgical margins were positive in 28 (46%) patients, and in 64% (18/28), the tumour extended to the deep margin, while the mucosal margin was positive in 6 patients (21%), and the location of the positive margin was unknown in 4. The sub-site breakdown for the positive margins were as follows: 20 (71%) were at the base of tongue, 7 (25%) were at the soft palate, and 1 (4%) was at the tonsils. Thirty-nine patients (64%) underwent concomitant neck dissection, including 21 who received END for N0 disease (54%) and 18 who underwent TND (46%). Seventeen neck dissections (44%) were comprehensive, 17 (44%) were selective, and 5 (13%) were bilateral. Thirty-five patients (57%) received adjuvant post-operative radiotherapy. Multivariate analyses demonstrated that the clinical T stage, anatomic sub-sites, and margin status were independent predictors of OS, and advanced clinical stage and a positive margin status were predictive of loco-regional recurrence. END should be considered for patients with an advanced T stage, as well as any T stage of ACC, MEC, adenocarcinoma (not otherwise specified) and the more recently identified MMSGC subtype polymorphous low-grade adenocarcinoma with involvement of the tongue [14, 18].

The introduction of transoral robotic surgery, with its improved three-dimensional imaging and manual dexterity, is likely to achieve a high rate of positive excision margins, and consequently, evidence of improved survival is being awaited. Early results on the use of transoral robotic surgery for MMSGC have demonstrated that it is feasible and produces low morbidity and good functional and oncologic results [19].

Nose: Paranasal Sinuses and Nasopharynx. ACC is most frequently found in the nasal cavity, paranasal sinuses and nasopharynx, followed by MEC, and these two sutypes account for >85% of cases [20]. Due to submucosal infiltration and bone destruction, the majority of MMSGN patients present at an advanced stage, resulting in difficulties with resectability and a worse prognosis compared with cancers of any other area of the head and neck. In 163 patients treated at Tata Memorial Hospital, Mumbai [20], the 5-year DFS and OS rates were 48.3 and 83.3%, respectively. Tumour grade, nodal status, and adjuvant radiotherapy were significant predictors of DFS, as determined by multivariate analysis. Frequent local recurrences and early perineural and haematogenous spread were observed, and 50% of the patients presented with distant metastases.

Recent progress in surgery with endoscopic endonasal resection (EER) has permitted comprehensive local surgical excision while allowing for good reconstruction that facilitates adequate post-operative radiotherapy. EER was initially used in select cases of nasal malignancy, but the use of this procedure has since expanded, and its use in combination with frontal craniotomy (cranio-endoscopic resection) has broadened the surgical indications. Developments in surgical techniques and instruments have enabled access to 'hidden areas,' such as the frontal sinus and posterior wall of the frontal sinus. For example, using frontal drill-out, endoscopic medial maxillectomy permits access to the entire maxillary sinus and the infratemporal fossa [21]. Endoscopic endonasal nasopharyngectomy has demonstrated feasibility and minimal morbidity in select cases of malignancy, and the local control rates reported are comparable to those of conventional procedures [22]. The inclusion of robotic instruments has been advocated to allow for more extensive tumours to be surgically approached with curative intent [23]. While dramatic and impressive results have been achieved with the use of EER, the published outcomes thus far have been reported by extremely experienced endoscopic surgeons. Such work involving the full spectrum of the skull base and sinonasal cancer surgery must include the gold standard techniques of open craniofacial resection and transfacial/transcranial approaches [24].

Larynx. Laryngeal salivary gland carcinomas are extremely rare, accounting for <1% of all malignant tumours in this region. The most common types are ACC, MEC and ADNOS [25, 26]. The largest number of reported cases have been located in the subglottis, and locations of the supraglottis in the false cords, aryepiglottic folds and caudal aspect of the epiglottis have also been reported. In the glottis, these tumours are located in the floor of the sinus of Morgagni and the subglottic surface of the anterior commissure. They most commonly present as a submucosal mass, and because they spread in a submucosal manner, they present late as large tumours. Local symptoms, such as hoarseness and swallowing problems, occur due to a mass effect, whereas breathing problems are most often associated with a subglottic lesion. Partial laryngeal surgery is possible in a small number of select cases, but total laryngectomy followed by post-operative radiotherapy is generally required because of the extensive submucosal spread [25, 27]. Two patients who refused surgery were treated with combined chemoradiotherapy for laryngeal ACC. Both were reported to be alive with locoregional control and functional larynges, one at 112+ months, with pulmonary metastasis (transglottic T4) at 54 months, and the other (glottic T3) who was malignant tumour free (local, regional and distant) at 60+ months [28]. Thus, patients with MMSGN who refuse total laryngectomy may be treated with chemoradiotherapy with curative intent, depending on the patient's choice and his or her suitability for such non-surgical management.

References

1 Zdabowski R, Dias FL, Barbosa MM, et al: Sublingual gland tumours: clinical, pathologic, and therapeutic analysis of 13 patients treated in a single unit. Head Neck 2011;33:476–481.
2 Yu T, Gao Q-H, Wang X-Y, et al: Malignant sublingual gland tumours: a retrospective clinicopathologic study of 28 cases. Oncology 2007;72:39–44.
3 Sun G, Yang X, Tang E, et al: The treatment of sublingual gland tumours. Oral Maxillofac Surg 2010;39:863–868.
4 Spiro RH: Treating tumours of the sublingual glands, including a useful technique for repair of the floor of mouth after resection. Am J Surg 1995;170:457–460.
5 Vander Poorten V, Hunt J, Bradley PJ, et al: Recent trends in the management of minor salivary gland carcinoma. Head Neck 2014;36:444–455.
6 Speight PM: Update on diagnostic difficulties in lesions of the minor salivary glands. Head Neck Pathol 2007;1:55–60.
7 Turk AT, Wenig BM: Pitfalls in the biopsy diagnosis of intraoral minor salivary gland neoplasms: diagnostic considerations and recommended approach. Adv Anat Pathol 2014;21:1–11.
8 Matsuzaki H, Yanagi Y, Hara M, et al: Diagnostic value of dynamic contrast-enhanced MRI for submucosal palatal tumors. Eur J Radiol 2012;81:3306–3312.
9 Sharma P, Jain TK, Singh H, et al: Utility of (18)F-FDG PET-CT in staging and restaging of patients with malignant salivary gland tumours: a single-institutional experience. Nucl Med Commun 2013;34:211–219.
10 Vander Poorten V, Balm AJ, Hilgers FJ, et al: Stage as major long term outcome predictor in minor salivary gland carcinoma. Cancer 2000;89:1195–204.
11 Terhaard CH: Postoperative and primary radiotherapy for salivary gland carcinomas; indications, techniques and results. Int J Radiat Oncol Biol Phys 2007;69(2 suppl):S52–S53.
12 Lloyd S, Yu JB, Ross DA, et al: A prognostic index for predicting lymph node metastasis in minor salivary gland cancer. Int J Radiat Oncol Biol Phys 2010;76:169–175.
13 Chen AM, Garcia J, Lee NY, et al: Patterns of nodal relapse after surgery and postoperative radiation therapy for carcinomas of the major and minor salivary glands: what is the role of elective neck irradiation? Int J Radiat Oncol Biol Phys 2007;67:988–994.
14 Li Q, Zhang XR, Liu XK, et al: Long-term outcome of minor salivary gland carcinoma of the hard palate. Oral Oncol 2011;48:456–462.
15 Kakarala K, Bhattacharyya N: Survival in oral cavity minor salivary gland carcinoma. Otolaryngol Head Neck Surg 2010;143:122–126.
16 Vaidya AD, Pantyaidya GH, Metgudmath R, et al: Minor salivary gland tumors of the oral cavity: a case series with review of the literature. J Can Res Therapeut 2010;8(suppl 2):S111–S115.

17 Iyer GN, Kim L, Nixon IJ, et al: Factors predicting outcome in malignant minor salivary gland tumors of the oropharynx. Arch Otolaryngol Head Neck 2010; 136:1240–1247.
18 Seethala RR, Johnson JT, Barnes L, et al: Polymorphous low-grade adenocarcinoma. Arch Otolaryngol Head Neck Surg 2010;136:385–392.
19 Villanueva NL, de Almeida JR, Sikora AG, et al: Transoral robotic surgery for the management of oropharyngeal minor salivary gland tumors. Head Neck 2014;36:28–33.
20 Pantvaidya GH, Vaidya AD, Metgudmath R, et al: Minor salivary gland tumors of the sinonasal region: results of a retrospective analysis with review of the literature. Head Neck 2012;34:1704–1710.
21 Lund VJ, Stammberger H, Nicolai P, et al: European position paper on endoscopic management of tumours of the nose, paranasal sinuses and skull base. Rhinology Suppl 2010;1:1–143.
22 Castelnuovo P, Nicolai P, Zurri-Zanoni M, et al: Endoscopic endonasal nasopharyngectomy in selected cancers. Otolaryngol Head Neck Surg 2014;149:424–430.
23 Carrau RL, Prevedello DM, de Lara D, et al: Combined transoral robotic surgery and endoscopic endonasal approach for the resection of extensive malignancies of the skull base. Head Neck 2013;35:351–358.
24 Castelnuovo P, Turri-Zanoni M, Battaglia P, et al: Endoscopic endonasal approaches for malignant tumours involving the skull base. Curr Otorhinolaryngol Rep 2013;1:197–205.
25 Ganly I, Patel SG, Coleman M, et al: Malignant minor salivary gland tumors of the larynx. Arch Otolaryngol Head Neck Surg 2006;132:767–770.
26 Nielsen TK, Bjorndal K, Krogdahl A, et al: Salivary gland carcinomas of the larynx: a national study in Denmark. Auris Nasus Larynx 2012;39:611–614.
27 Zhang M, Li KN, Li C, et al: Malignant salivary gland carcinomas of the larynx. ORL J Otorhinolaryngol Relat Spec 2014; 76:222–226.
28 Misiukiewicz KJ, Camille N, Tishler R, et al: Organ preservation for adenoid cystic carcinoma of the larynx. Oncologist 2013;18:579–583.

Patrick J. Bradley, MBA, FRCS
Emeritus Honorary Professor at the School of Medicine, The University of Nottingham
Honorary Consultant Head and Neck Oncological Surgeon, Nottingham University Hospitals
Queens Medical Centre Campus, Derby Road
Nottingham NG7 2UH (UK)

10 Chartwell Grove, Mapperley Plains
Nottingham NG3 5RD (UK)
E-Mail pjbradley@zoo.co.uk

Facial Reconstruction and Rehabilitation

Orlando Guntinas-Lichius[a] · Dane J. Genther[b] · Patrick J. Byrne[b,c]

[a]Department of Otorhinolaryngology, Jena University Department, Jena, Germany; Departments of [b]Otolaryngology – Head and Neck Surgery and [c]Dermatology, The Johns Hopkins University School of Medicine, Baltimore, Md., USA

Abstract

Extracranial infiltration of the facial nerve by salivary gland tumors is the most frequent cause of facial palsy secondary to malignancy. Nevertheless, facial palsy related to salivary gland cancer is uncommon. Therefore, reconstructive facial reanimation surgery is not a routine undertaking for most head and neck surgeons. The primary aims of facial reanimation are to restore tone, symmetry, and movement to the paralyzed face. Such restoration should improve the patient's objective motor function and subjective quality of life. The surgical procedures for facial reanimation rely heavily on long-established techniques, but many advances and improvements have been made in recent years. In the past, published experiences on strategies for optimizing functional outcomes in facial paralysis patients were primarily based on small case series and described a wide variety of surgical techniques. However, in the recent years, larger series have been published from high-volume centers with significant and specialized experience in surgical and nonsurgical reanimation of the paralyzed face that have informed modern treatment. This chapter reviews the most important diagnostic methods used for the evaluation of facial paralysis to optimize the planning of each individual's treatment and discusses surgical and nonsurgical techniques for facial rehabilitation based on the contemporary literature.

© 2016 S. Karger AG, Basel

Introduction

Facial paralysis is caused by a tumor invading the facial nerve along its course from the brainstem to the facial musculature and affects approximately 5% of patients with salivary gland tumors [1]. The most frequent etiology among this subset of patients is the topic of this chapter: extracranial infiltration of the facial nerve by a malignant parotid tumor. The incidence of facial palsy in individuals with parotid cancer at the time of presentation is 12–15% [2]. For resectable salivary gland cancer with preoperative paralysis, the treatment of choice is radical parotidectomy with resection of the involved segment of the facial nerve [3]. The degree to which the facial nerve is infiltrated varies from partial involvement of one branch to complete involvement of the main trunk. If feasible from a technical and oncologic standpoint, the facial nerve should be reconstructed immediately following tumor resection. Another, albeit less common scenario, is resection and reconstruction of the marginal mandibular branch in the case of submandibular gland cancer. Finally, patients treated for salivary gland cancer who have long-term facial palsy, either because primary nerve reconstruction was not performed following resection or because full recovery of postoperative palsy was expected but not realized, may desire secondary facial reanimation.

When confronted with facial weakness secondary to a salivary gland neoplasm that requires surgical correction, achievement of an optimal patient outcome requires experience in a variety of facial reanimation techniques [1, 4, 5]. The exact techniques and procedures to be employed depend on many factors, including the site of the lesion, extent of the current and/or expected palsy, viability of the proximal facial nerve stump, denervation time, and sensory function of the trigeminal nerve. Additionally, the surgeon must consider the patient's age, medical comorbidities, and wishes. Adept assessment of these factors allows for determination of an individualized approach to repair and reanimation. Fortunately, facial analysis and diagnostics have undergone some important advances in recent years with regard to the establishment of objective tools for the assessment of facial weakness that have largely replaced traditional subjective facial nerve grading [6, 7]. The goal of these objective tools is to more accurately and reliably evaluate facial nerve function prior to and after facial nerve reconstruction.

Preoperative Assessment

In accordance with oncologic principles, reconstruction of the facial nerve is subordinate to primary treatment of the salivary gland neoplasm. The goal of surgical reconstruction of the facial nerve is to restore function of the mimic musculature. Ideally, this includes reestablishment of the resting tone of all mimic muscles and restoration of frontal frowning with lifting of the eyebrows, closure of the eyes, symmetry of the nasolabial fold, and the ability to laugh symmetrically. Practically, restoration of eye closure and smile reanimation receive the highest priority in facial reanimation surgery [8]. During preoperative assessment, the evaluator must confirm the facial nerve lesion as the etiology of the facial palsy, confirm the irreversibility of the damage, and determine the exact extent of the lesion. The most useful diagnostic tools are the patient's history, clinical examination, electrodiagnostics, and in some cases, imaging. More recently, assessment of an individual patient's concerns using patient-reported outcome measures (PROMs) has gained increased attention [9].

Electrophysiological Profile

In contrast to acute facial palsy, electroneurography does not play a significant role in patients with a proven facial nerve lesion due to a salivary gland neoplasm; however, electromyography (EMG) does play a central role in the electrophysiological evaluation of such patients. Each facial muscle of interest and its related facial nerve branch can be assessed by EMG, establishing an individual electrophysiological profile [10]. Acute facial nerve damage due to a malignant tumor or due to tumor resection causes pathologic alterations in the innervation of target facial muscles [11]. Chronic facial nerve damage in long-term palsy leads to muscular atrophy, which can be demonstrated by alterations of the insertion potentials during needle EMG. EMG can be used to monitor the process of nerve regeneration after facial nerve reconstruction and precedes clinical regeneration [12]. Finally, EMG allows for quantification of the functional outcome after facial reanimation [13, 14], as well as quantification of defective healing, namely synkinesis and dyskinesis, after recovery is complete [12].

Imaging

Preoperative facial palsy from salivary gland cancer can be the result of tumor invasion or perineural spread. MRI is the method of choice for localization of the site of the lesion and for determination of the extent of perineural spread, providing useful information for determining appropriate surgical resection [15]. Additionally, MRI can be used to evaluate the degree of atrophy of the facial musculature in cases of long-term denervation and to monitor facial muscle growth after reinnervation following facial nerve surgery [16].

Recently, protocols have been established for standardized quantitative ultrasound investigations of most major facial muscles [17, 18]. This technology can be used for regional and quantitative evaluation of facial muscles in patients with facial palsy [19], and after facial nerve reconstruction, serial ultrasound examinations can be used to monitor regeneration of the facial muscles, ideally showing progressive facial muscle regrowth [20].

Evaluation of Facial Motor Function

Clinicians need a uniform, objective, accurate, reliable, simple, and sensitive facial grading system to characterize facial function before and after facial reanimation surgery. Many efforts have been made in recent years to develop automated, computer-based systems. The Facial Nerve Center in Boston routinely uses the validated Facial Assessment by Computer Evaluation software (FACE-gram) to evaluate facial movement using images from before and after facial nerve grafting and free flap reconstructions [6, 21, 22]. At the University of Pittsburgh, an automated facial image analysis system has been recently developed and validated [23]. This system has been used to objectively evaluate facial muscle activity in patients with partial facial palsy who received a fascial sling to enhance facial movement [24]. Specific and functionally relevant facial expression tasks, such as puckering and blowing, can be quantified [25]. Additionally, computerized 3-D evaluation software has been developed through analysis of standardized photographs to precisely assess deficits in eyelid function in facial palsy [26]. Recently, it has been shown that automated analysis of Facial Action Coding – primarily developed for dynamic analysis of facial expression in psychological research – represents a simple tool for analysis of facial motor function in patients with facial palsy in a clinical setting [27].

Evaluation of Facial Nonmotor Function

In addition to be being a movement disorder, facial palsy can result in significant emotional distress and communicative dysfunction in some patients [28]. Both the functional and psychological impairments lead to a decreased quality of life [29]. In addition to the tools used to assess general quality of life, several relevant PROMs now exist for facial paralysis, such as the Facial Clinimetric Evaluation (FaCE) Scale, Nasal Obstruction Symptom Evaluation (NOSE) Scale, and Synkinesis Assessment Questionnaire [30–32]. Such measures are essential for completing patient assessments and should be employed by all healthcare providers who treat facial paralysis [9].

Several studies have demonstrated an association between the success of facial reanimation surgery and improved quality of life. A recent prospective study reported that improvement in the nasal obstruction-related quality-of-life NOSE score following fascia lata sling placement is significantly correlated with correction of nasal valve compromise [33]. In another recent study, the FaCE score and FACE-gram were used together to show quantitative improvements in quality of life and facial motor function after free gracilis muscle transfer for smile restoration [22]. There is also evidence that facial reanimation surgery is associated with improved facial expression of emotion [34]. Furthermore, facial reanimation has been demonstrated to increase attractiveness and decrease the negative facial perception of patients with facial paralysis [35].

Selection of an Individualized Facial Rehabilitation Treatment Plan

The basis for selection of the rehabilitation technique(s) to be used includes the extent of the extratemporal lesion and the duration of paralysis. According to these two parameters, all surgical rehabilitation techniques can be divided into three main categories (table 1): (a) reconstruction of the facial nerve at the time of salivary gland cancer resection; (b) immediate/early reanimation when the facial nerve cannot be directly reconstructed; and (c) delayed or late reconstruction [4]. Immediate or early reconstruction occurs within the first 2 months after facial nerve injury and affords the best opportunity for functional recovery. Late reconstruction is performed at 12–18 months after onset of the palsy. Beyond 12 months of denervation time, irreversible atrophy and fibrosis of the facial musculature sets in if no regeneration occurs. Additionally, if spontaneous but functionally insufficient regeneration does occur during this 12-month or longer period of time, then defective healing in the form of synkinesia and dyskinesia has likely reached its final stage. Patients with durations of palsy of between 2 and 12 months fall into a less certain category. These patients are difficult to categorize and must be considered individually after complete diagnostic examination.

In general, if reconstruction of the facial nerve (first line) is not possible or if there is not a chance for meaningful recovery because the facial muscles are irreversibly atrophied, then muscle transposition or free muscle transfer (second line) should be considered, as this allows for dynamic reanimation of the paralyzed face. Static reconstruction with slings and other repositioning techniques is third line in the hierarchy of reconstruction techniques for facial reanimation. In reality, multiple procedures from one, two, or all three of these technical categories are performed concomitantly.

Surgical Rehabilitation

Most of the surgical techniques for facial rehabilitation are well known and long established. This chapter focuses on the innovations and clinical studies of recent years. To optimize facial reanimation for each individual, the surgeon should be comfortable with a wide variety of standard techniques and an understanding of how these techniques can be combined. Additionally, overall care for patients with facial paralysis should be administered by a multidisciplinary team to maximize outcomes [36, 37].

Facial Nerve Suture and Grafting Techniques

The hypoglossal-facial jump nerve suture is the most often used procedure for reanimation if the proximal facial nerve is unavailable due to its resection with the primary tumor. If adequate length in unavailable for end-to-side anastomosis of the distal facial nerve or its branches to the hypoglossal nerve, then an interposition graft is used. However, Beutner et al. [38] have shown that the jump technique is sometimes feasible without a graft if the facial nerve is adequately mobilized from the fallopian canal following mastoidectomy.

For complex defects of the extratemporal facial fan, the hypoglossal jump technique can used to reanimate the lower face in combination with cross-facial nerve grafting to reanimate the upper face. An advantage conferred by reanimation of the upper and lower face using different neural afferents is the avoidance of synkinesis between the upper and lower face following reinnervation [39].

The masseteric nerve is becoming more popular as another valuable source for reanimation of the paralyzed face [40, 41]. This nerve can be used

Table 1. Plan by stages for facial reanimation*

	Surgical method	Comments
A.	Early reconstruction of extratemporal lesion	
	Step I:	
	A.1 Primary direct nerve suture	A.1 peripheral lesion ≤1 cm, avoid any tension
	A.2 Interpositional graft	A.2 peripheral lesions >1 cm, simple segmental defect
	A.3 Upper lid weight	A.3 better than tarsorrhaphy
	Step II:	
	A.3 Adjuvant measures	
B.	Early reanimation when facial nerve is not available (see above)	
	Step I:	
	B.1 Hypoglossal-facial jump anastomosis	B.1 better than classical hypoglossal-facial anastomosis; proximal facial nerve stump not available or proximal facial nerve stump available but peripheral lesions >1 cm, complex segmental defect
	B.2 Upper lid weight	
	B.3 Combined approach	B.3 complex segmental defect
	B.4 Massetric nerve transposition	B.4 complex segmental defect
	B.5 Cross-face nerve suture	
	B.6 Temporalis muscle transfer	B.6 better than masseter muscle transfer
	B.7 Digastric muscle transfer	
	B.8 Sling plasty	
	Step II:	
	B.9 Eyebrow lift	B.10 in case of brow ptosis
	B.10 Rhinoplasty	
	B.11 Rhytidectomy	B.12 in case of nasal asymmetry
	B.12 Botulinum toxin, myectomy	B.13 in case of cheek or chin ptosis
C.	Delayed/late reconstruction	
	Step I:	
	Mimic musculature existing:	
	C.1 Hypoglossal-facial jump anastomosis	C.1 Hypoglossal nerve: better than any other donor nerve
	C.2 Upper lid weight	
	C.3 Cross-face nerve suture	
	C.4 Massetric nerve transposition	
	Mimic musculature not existing but nerve supply existing:	
	C.5 Microvascular muscle transfer	C.5 Best choice/experience for congenital lesions
	C.6 Temporalis muscle transfer	
	Mimic musculature not existing and nerve supply not existing:	
	C.7 Sling plasty	C.7 Use palmaris longus tendon or fascia lata
	Step II:	
	C.8 Eyebrow lift	
	C.9 Rhinoplasty	
	C.10 Rhytidectomy	
	C.11 Botulinum toxin, myectomy	C.11 Correction of defective healing or facial asymmetry on lesioned and healthy sides

Adapted from [4, 70].

like the hypoglossal nerve for direct motor neurotization [42], in conjunction with cross-facial grafting [43], or for innervation of a free muscle flap reconstruction [44].

If a nerve graft is required, the sural or great auricular nerve can be harvested for an autologous graft; however, if these nerves are unavailable or undesirable, then cadaveric nerves can be

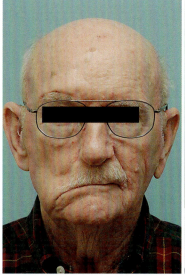
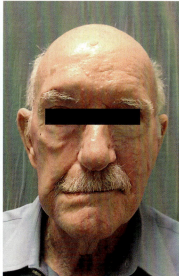

Fig. 1. Improved facial symmetry following temporalis tendon transfer. Photos provided by senior author (P.J.B.).

used. Motor nerve allografts are now commercially available for use for gaps of up to 5 cm [45]. Additionally, bioabsorbable conduits can be used in select cases with success [46].

Muscle and Tendon Transposition
An advantage of local functional muscle transfer that makes it a particularly attractive technique is that the effect is immediate, which is not the case with reinnervation procedures that require time for nerve regrowth. The temporalis muscle is commonly accepted as the first choice for functional muscle transfer for restoration of symmetry of the upper lip and oral commissure [47]. The classical technique of temporalis muscle transfer involves transposition of the superior aspect of a strip of mobilized temporalis muscle to the upper lip; however, this often results in a depression at the temple and a bulge over the zygomatic arch. Because of these disadvantages, the classical procedure has since been largely replaced by temporalis tendon transfer (T3). This procedure offers improved facial symmetry at rest (fig. 1), as well as lip excursion of the previously paralyzed hemiface with smiling. Temporalis tendon transfer is performed in an orthodromic manner in which the distal temporalis tendon is detached from the coronoid of the mandible and inset into the modiolus near the oral commissure [48]. Recently, a minimally invasive transoral technique for performance of the T3 procedure has been described [49]. Intraoperative electrical stimulation of the released temporalis muscle tendon unit can simulate excursion of the temporalis muscle prior to inset. This intraoperative technique allows for the surgeon to determine the ideal tension that the muscle tendon unit should be under at the time of inset to maximize force generation and excursion [50]. Temporalis tendon transfer is also an option for secondary surgery after radiation therapy; however, irradiated patients show less excursion following the procedure on average compared to nonirradiated patients [51].

An infrequently used but viable second choice for local muscle transfer is the masseter muscle. The esthetic results and degree of dynamic movement following masseter muscle transfer are inferior to T3. To overcome some of these limitations, a staged, split masseter muscle transfer in combi-

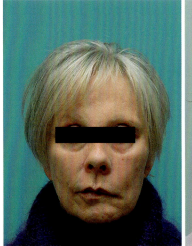
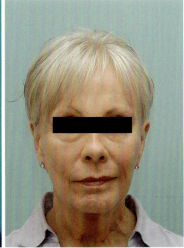

Fig. 2. Improved facial symmetry at rest after gracilis microneurovascular free tissue transfer. Photos provided by senior author (P.J.B.).

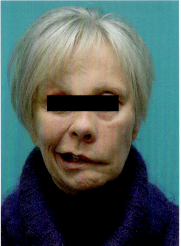
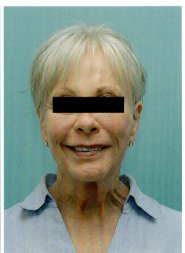

Fig. 3. Improved facial symmetry with smiling after gracilis microneurovascular free tissue transfer. Photos provided by senior author (P.J.B.).

nation with a hemiorial fascia lata graft to act as an anchor for reinforcement has been reported [52].

Free Muscle Transfer

Most experience with free muscle transfer has been gained through treatment of patients with congenital facial palsy. However, in recent years, an increasing number of studies have reported on the use of free muscle transfer in acquired palsy and after head and neck cancer surgery [53]. An important advantage of free functional muscle transfer is the ability to couple movement of the paralyzed face with that of the normally functioning contralateral face using cross-facial nerve grafting [45]. Additionally, some patients may not be candidates for local muscle transfer due to previous surgery, trauma, or other cranial nerve deficits.

The gracilis muscle transfer is the most widely described free functional muscle transfer technique for facial reanimation and is now the preferred technique for most patients [45, 54]. It is a

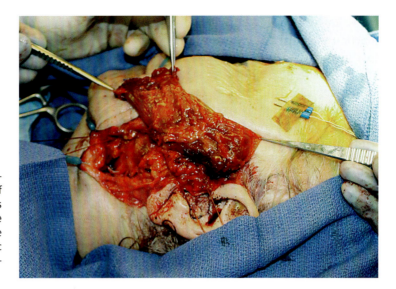

Fig. 4. Ideal vector for inset of gracilis muscle free flap. The distal end of the flap is inset into the soft tissues near the oral commissure, and the proximal end is suspended from the inferior aspect of the zygomatic arch. Photo provided by senior author (P.J.B.).

highly reliable technique that offers the ability to achieve near symmetry at rest (fig. 2) and, if a cross-facial nerve graft is used, a nearly symmetric spontaneous smile (fig. 3).

Other muscles that can be used include the rectus abdominus, pectoralis minor, latissimus dorsi, and sternohyoid; however, these have disadvantages that make them less desirable for use in facial reanimation surgery. The gracilis muscle can be harvested with minimal donor site morbidity using a two-team approach, and dissection of the flap is straightforward with consistent anatomy. Additionally, the neurovascular pedicle can be easily traced and harvested with adequate length for vascular and neural anastomoses [54]. To provide the most natural-appearing vector of pull for support of the face at rest and during smiling, the flap is inset into the soft tissues near the oral commissure and suspended from the inferior border of the zygomatic arch (fig. 4).

To provide mimetic movement, the gracilis is often performed as a staged procedure 8–12 months after cross-facial nerve grafting. However, the masseteric nerve can be used in some patients, allowing for less down time between muscle transfer and dynamic action. Additionally, dual innervation via cross-facial nerve grafting and masseteric nerve transfer has shown promise for mimetic movement with increased contractile power [54]. Hontanilla et al. [44] used an automatic optical motion system to compare the gracilis muscle transfer neurotized to a cross-facial nerve graft versus the masseteric nerve in 47 patients, and in another study of 46 patients, versus neurotization with the hemihypoglossal nerve [40]. In their hands, neurotization with the masseteric nerve provided better symmetry and a higher degree of recovery than cross-facial nerve grafting or hemihypoglossal nerve neurotization alone. The masseteric nerve can also be used in patients in a salvage procedure after failed cross-facial nerve grafting [55]. However, the use of the masseteric nerve has the potential disadvantage of causing involuntary and disturbing facial movements during mastication, a complication that is common in masseter muscle transfer procedures [56].

Sling Techniques
Static sling procedures for the midface are indicated for patients who are not candidates for nerve reconstruction or muscle transfer proce-

dures; however, they are often performed in conjunction with these dynamic procedures to further improve certain aspects of facial symmetry [24, 57]. Current methods for static slings are rather simplistic. Normally, the fascia lata or palmaris tendon is used to suspend the modiolus from the zygomatic arch or the temporalis fascia, with further extension to the midline of the upper and lower lips in a single vector [58]. Recently, a multivectorial approach using the palmaris tendon divided into 3 slips has been described, allowing for a well-defined nasolabial fold and sufficient cheek tone [58]. Nevertheless, compared with dynamic reconstruction techniques, the results have been limited. If an autograft cannot be used, then there are several allografts available, but there is no consensus regarding the optimal allogenic material [59].

Restoration of Eye Closure
Restoration of complete eye closure is necessary for adequate ocular lubrication and corneal protection and is one of the major focuses of functional facial reanimation surgery. Static upper eyelid procedures, including upper eyelid loading and tarsorrhaphy, are frequently used individually or as adjuvant treatments to nerve grafting techniques to assist with eye closure. Implantation of a platinum chain for upper eyelid loading is a mainstay of treatment for incomplete eye closure. Platinum chains consistently afford good functional results and improved quality of life [60], and their placement can be performed quickly in a clinic setting with minimal morbidity and cost. Schrom et al. [61] compared patient satisfaction after upper eyelid loading with either gold or platinum implants and have found that platinum chain use is associated with greater satisfaction.

Whereas reanimation of the upper eyelid is frequently performed, adjunctive lower eyelid procedures are often not performed despite their potential for providing additional improvement [62]. Treatment of paralytic ectropion can be performed through a variety of lateral canthoplasty techniques or lateral canthal tendon reinsertion [63]. Procedures to correct medial ectropion can be performed individually or in conjunction with lateral lid procedures in the case of generalized ectropion, and they include medial canthal tendon tightening, medial canthal tendon resuspension, and the medial spindle procedure. Lower eyelid laxity, which is common in elderly patients, can be treated with lid shortening procedures or spacer grafts [64].

Nonsurgical Rehabilitation

Chemodenervation with the botulinum toxin is an established nonsurgical therapy for postparetic or postsurgical synkinesis or hyperkinesis. Its effectiveness has been proven in several clinical studies using subjective scoring tools, and recently, its effectiveness at reducing synkinesis has been proven by objective image analysis [65]. Another study has used the FaCE scale to show that use of the botulinum toxin for synkinesis decreases the degrees of impairment and disability perceived by the patient [66].

Literature analyzing the role of adjuvant physiotherapy after facial reanimation surgery is currently lacking [67]. The majority of larger series have evaluated patients with Bell's palsy, and few have assessed patients after facial reanimation surgery [68]. However, a recent report concluded that EMG biofeedback therapy might be helpful in patients with chronic facial synkinesis [69].

Conclusion

The primary aims of facial reanimation surgery are to restore tone, symmetry, and movement to the paralyzed face. There are many reliable techniques available to fulfill these aims, and treatment plans can be tailored to each individual patient's situation and wishes. Due to the relative rarity of the disease, facial rehabilitation is best performed at a dedicated center with a multidisciplinary team. It is incumbent upon this team to demonstrate the results of facial reanimation surgery using objective measures and PROMs. An objective evaluation of this nature is the only way to standardize facial reanimation surgery and should be applied whenever possible. It is also important to remember that facial rehabilitation involves not only surgical but also nonsurgical practices. Adjuvant physiotherapy with facial retraining therapy and the selective use of chemodenervation agents are vital adjuncts to surgical facial reanimation. Ultimately, providers cannot forget that facial paralysis is not only a movement disorder but also a potentially disabling condition with social, emotional, and communicative ramifications. These psychological aspects must also be addressed as part of the multidisciplinary therapy of patients with facial paralysis.

References

1 Guntinas-Lichius O: The facial nerve in the presence of a head and neck neoplasm: assessment and outcome after surgical management. Curr Opin Otolaryngol Head Neck Surg 2004;12:133–141.
2 O'Brien CJ, Adams JR: Surgical management of the facial nerve in presence of malignancy about the face. Curr Opin Otolaryngol Head Neck Surg 2001;9:90–94.
3 Spiro RH: Changing trends in the management of salivary tumors. Semin Surg Oncol 1995;11:240–245.
4 Volk GF, Pantel M, Guntinas-Lichius O: Modern concepts in facial nerve reconstruction. Head Face Med 2011;6:25.
5 Chan JY, Byrne PJ: Management of facial paralysis in the 21st century. Facial Plast Surg 2011;27:346–357.
6 Hadlock TA, Urban LS: Toward a universal, automated facial measurement tool in facial reanimation. Arch Facial Plast Surg 2012;14:277–282.
7 Fattah A, Gurusinghe A, Gavilan J, Hadlock T, Marcus J, Marres H, Nduka C, Slattery W, Snyder-Warwik A; Sir Charles Bell Society: Facial nerve grading instruments: systematic review of the literature and suggestion for uniformity. Plast Reconstr Surg 2015;135:569–579.
8 Hohman MH, Hadlock TA: Etiology, diagnosis, and management of facial palsy: 2,000 patients at a facial nerve center. Laryngoscope 2014;124:E283–E293.
9 Bhama P, Gliklich RE, Weinberg JS, Hadlock TA, Lindsay RW: Optimizing total facial nerve patient management for effective clinical outcomes research. JAMA Facial Plast Surg 2014;16:9–14.
10 Kennelly KD: Electrodiagnostic approach to cranial neuropathies. Neurol Clin 2012;30:661–684.
11 Grosheva M, Wittekindt C, Guntinas-Lichius O: Prognostic value of electroneurography and electromyography in facial palsy. Laryngoscope 2008;118:394–397.
12 Guntinas-Lichius O, Streppel M, Stennert E: Postoperative functional evaluation of different reanimation techniques for facial nerve repair. Am J Surg 2006;191:61–67.
13 Michaelidou M, Herceg M, Schuhfried O, Tzou CH, Pona I, Hold A, Mittlbock M, Paternostro-Sluga T, Frey M: Correlation of functional recovery with the course of electrophysiological parameters after free muscle transfer for reconstruction of the smile in irreversible facial palsy. Muscle Nerve 2011;44:741–748.
14 Sajjadian A, Song AY, Khorsandi CA, Deleyiannis FW, VanSwearingen JM, Henkelmann TC, Hui K, Manders EK: One-stage reanimation of the paralyzed face using the rectus abdominis neurovascular free flap. Plast Reconstr Surg 2006;117:1553–1559.
15 Gandhi MR, Panizza B, Kennedy D: Detecting and defining the anatomic extent of large nerve perineural spread of malignancy: comparing 'targeted' MRI with the histologic findings following surgery. Head Neck 2011;33:469–475.
16 Volk GF, Karamyan I, Klingner CM, Reichenbach JR, Guntinas-Lichius O: Quantitative magnetic resonance imaging volumetry of facial muscles in healthy patients with facial palsy. Plast Reconstr Surg Glob Open 2014;2:e173.
17 Volk GF, Wystub N, Pohlmann M, Finkensieper M, Chalmers HJ, Guntinas-Lichius O: Quantitative ultrasonography of facial muscles. Muscle Nerve 2013;47:878–883.
18 Volk GF, Sauer M, Pohlmann M, Sauer M, Guntinas-Lichius O: Reference values for dynamic facial muscle ultrasonography in adults. Muscle Nerve 2014;50:348–357.
19 Volk GF, Pohlmann M, Finkensieper M, Chalmers HJ, Guntinas-Lichius O: 3D-ultrasonography for evaluation of facial muscles in patients with chronic facial palsy or defective healing: a pilot study. BMC Ear Nose Throat Disord 2014;14:4.
20 Volk GF, Pohlmann M, Sauer M, Finkensieper M, Guntinas-Lichius O: Quantitative ultrasonography of facial muscles in patients with chronic facial palsy. Muscle Nerve 2014;50:358–365.
21 Lee LN, Susarla SM, M HH, Henstrom DK, Cheney ML, Hadlock TA: A comparison of facial nerve grading systems. Ann Plast Surg 2013;70:313–316.

22 Lindsay RW, Bhama P, Hadlock TA: Quality-of-life improvement after free gracilis muscle transfer for smile restoration in patients with facial paralysis. JAMA Facial Plast Surg 2014;16:419–424.

23 Wachtman GS, Cohn JF, VanSwearingen JM, Manders EK: Automated tracking of facial features in patients with facial neuromuscular dysfunction. Plast Reconstr Surg 2001;107:1124–1133.

24 Deleyiannis FW, Askari M, Schmidt KL, Henkelmann TC, VanSwearingen JM, Manders EK: Muscle activity in the partially paralyzed face after placement of a fascial sling: a preliminary report. Ann Plast Surg 2005;55:449–455.

25 Denlinger RL, VanSwearingen JM, Cohn JF, Schmidt KL: Puckering and blowing facial expressions in people with facial movement disorders. Phys Ther 2008;88:909–915.

26 Marron Mendes V, Lasudry J, Vandermeeren L, de Fontaine S: Computerised 3D evaluation of the functional eyelid deficit in facial palsy. J Plast Reconstr Aesthet Surg 2014;67:178–182.

27 Haase D, Minnigerode L, Volk GF, Denzler J, Guntinas-Lichius O: Automated and objective action coding of facial expressions in patients with acute facial palsy. Eur Arch Otorhinolaryngol 2015;272:1259–1267.

28 Dobel C, Miltner WH, Witte OW, Volk GF, Guntinas-Lichius O: Emotional impact of facial palsy (in German). Laryngorhinootologie 2013;92:9–23.

29 Guntinas-Lichius O, Straesser A, Streppel M: Quality of life after facial nerve repair. Laryngoscope 2007;117:421–426.

30 Kahn JB, Gliklich RE, Boyev KP, Stewart MG, Metson RB, McKenna MJ: Validation of a patient-graded instrument for facial nerve paralysis: the face scale. Laryngoscope 2001;111:387–398.

31 Stewart MG, Witsell DL, Smith TL, Weaver EM, Yueh B, Hannley MT: Development and validation of the nasal obstruction symptom evaluation (NOSE) scale. Otolaryngol Head Neck Surg 2004;130:157–163.

32 Mehta RP, WernickRobinson M, Hadlock TA: Validation of the synkinesis assessment questionnaire. Laryngoscope 2007;117:923–926.

33 Lindsay RW, Bhama P, Hohman M, Hadlock TA: Prospective evaluation of quality-of-life improvement after correction of the alar base in the flaccidly paralyzed face. JAMA Facial Plast Surg 2015;17:108–112.

34 Dey JK, Ishii M, Boahene KD, Byrne PJ, Ishii LE: Facial reanimation surgery restores affect display. Otol Neurotol 2014;35:182–187.

35 Dey JK, Ishii M, Boahene KD, Byrne PJ, Ishii LE: Changing perception: facial reanimation surgery improves attractiveness and decreases negative facial perception. Laryngoscope 2014;124:84–90.

36 Boahene KD: Principles and biomechanics of muscle tendon unit transfer: application in temporalis muscle tendon transposition for smile improvement in facial paralysis. Laryngoscope 2013;123:350–355.

37 Bhama PK, Hadlock TA: Contemporary facial reanimation. Facial Plast Surg 2014;30:145–151.

38 Beutner D, Luers JC, Grosheva M: Hypoglossal-facial-jump-anastomosis without an interposition nerve graft. Laryngoscope 2013;123:2392–2396.

39 Volk GF, Pantel M, Streppel M, Guntinas-Lichius O: Reconstruction of complex peripheral facial nerve defects by a combined approach using facial nerve interpositional graft and hypoglossal-facial jump nerve suture. Laryngoscope 2011;121:2402–2405.

40 Hontanilla B, Marre D: Comparison of hemihypoglossal nerve versus masseteric nerve transpositions in the rehabilitation of short-term facial paralysis using the facial clima evaluating system. Plast Reconstr Surg 2012;130:662e–672e.

41 Collar RM, Byrne PJ, Boahene KD: The subzygomatic triangle: rapid, minimally invasive identification of the masseteric nerve for facial reanimation. Plast Reconstr Surg 2013;132:183–188.

42 Henstrom DK: Masseteric nerve use in facial reanimation. Curr Opin Otolaryngol Head Neck Surg 2014;22:284–290.

43 Bianchi B, Ferri A, Ferrari S, Copelli C, Magri A, Ferri T, Sesenna E: Cross-facial nerve graft and masseteric nerve cooptation for one-stage facial reanimation: principles, indications, and surgical procedure. Head Neck 2014;36:235–240.

44 Hontanilla B, Marre D, Cabello A: Facial reanimation with gracilis muscle transfer neurotized to cross-facial nerve graft versus masseteric nerve: a comparative study using the facial clima evaluating system. Plast Reconstr Surg 2013;131:1241–1252.

45 Boahene K: Reanimating the paralyzed face. F1000Prime Rep 2013;5:49.

46 Inada Y, Hosoi H, Yamashita A, Morimoto S, Tatsumi H, Notazawa S, Kanemaru S, Nakamura T: Regeneration of peripheral motor nerve gaps with a polyglycolic acid-collagen tube: technical case report. Neurosurgery 2007;61:E1105–E1107; discussion E1107.

47 Sidle DM, Simon P: State of the art in treatment of facial paralysis with temporalis tendon transfer. Curr Opin Otolaryngol Head Neck Surg 2013;21:358–364.

48 Byrne PJ, Kim M, Boahene K, Millar J, Moe K: Temporalis tendon transfer as part of a comprehensive approach to facial reanimation. Arch Facial Plast Surg 2007;9:234–241.

49 Boahene KD, Farrag TY, Ishii L, Byrne PJ: Minimally invasive temporalis tendon transposition. Arch Facial Plast Surg 2011;13:8–13.

50 Boahene KD, Ishii LE, Byrne PJ: In vivo excursion of the temporalis muscle-tendon unit using electrical stimulation: application in the design of smile restoration surgery following facial paralysis. JAMA Facial Plast Surg 2014;16:15–19.

51 Griffin GR, Abuzeid W, Vainshtein J, Kim JC: Outcomes following temporalis tendon transfer in irradiated patients. Arch Facial Plast Surg 2012;14:395–402.

52 Lesavoy MA, Fan KL, Goldberg AG, Dickinson BP, Herrera F: Facial reanimation by staged, split masseter muscle transfer. Ann Plast Surg 2014;73:33–38.

53 Terzis JK, Konofaos P: Experience with 60 adult patients with facial paralysis secondary to tumor extirpation. Plast Reconstr Surg 2012;130:51e–66e.

54 Revenaugh PC, Byrne PJ: Gracilis microneurovascular transfer for facial paralysis. Facial Plast Surg 2015;31:134–139.

55 Eisenhardt SU, Eisenhardt NA, Thiele JR, Stark GB, Bannasch H: Salvage procedures after failed facial reanimation surgery using the masseteric nerve as the motor nerve for free functional gracilis muscle transfer. JAMA Facial Plast Surg 2014;16:359–363.

56 Rozen S, Harrison B: Involuntary movement during mastication in patients with long-term facial paralysis reanimated with a partial gracilis free neuromuscular flap innervated by the masseteric nerve. Plast Reconstr Surg 2013; 132:110e–116e.

57 Leckenby JI, Harrison DH, Grobbelaar AO: Static support in the facial palsy patient: a case series of 51 patients using tensor fascia lata slings as the sole treatment for correcting the position of the mouth. J Plast Reconstr Aesthet Surg 2014;67:350–357.

58 Yoleri L, Gungor M, Usluer A, Celik D: Tension adjusted multivectorial static suspension with plantaris tendon in facial paralysis. J Craniofac Surg 2013; 24:896–899.

59 Ibrahim AM, Rabie AN, Kim PS, Medina M, Upton J, Lee BT, Lin SJ: Static treatment modalities in facial paralysis: a review. J Reconstr Microsurg 2013;29: 223–232.

60 Bianchi B, Ferri A, Leporati M, Ferrari S, Lanfranco D, Ferri T, Sesenna E: Upper eyelid platinum chain placement for treating paralytic lagophthalmos. J Craniomaxillofac Surg 2014;42:2045–2048.

61 Schrom T, Buchal A, Ganswindt S, Knipping S: Patient satisfaction after lid loading in facial palsy. Eur Arch Otorhinolaryngol 2009;266:1727–1731.

62 Loyo M, Jones D, Lee LN, Collar RM, Molendijk J, Boahene KD, Ishii LE, Byrne PJ: Treatment of the periocular complex in paralytic lagophthalmos. Ann Otol Rhinol Laryngol 2015;124:273–279.

63 Korteweg SF, Stenekes MW, van Zyl FE, Werker PM: Paralytic ectropion treatment with lateral periosteal flap canthoplasty and introduction of the ectropion severity score. Plast Reconstr Surg Glob Open 2014;2:e151.

64 Rahman I, Sadiq SA: Ophthalmic management of facial nerve palsy: a review. Surv Ophthalmol 2007;52:121–144.

65 Cecini M, Pavese C, Comelli M, Carlisi E, Sala V, Bejor M, Dalla Toffola E: Quantitative measurement of evolution of postparetic ocular synkinesis treated with botulinum toxin type a. Plast Reconstr Surg 2013;132:1255–1264.

66 Couch SM, Chundury RV, Holds JB: Subjective and objective outcome measures in the treatment of facial nerve synkinesis with onabotulinumtoxina (botox). Ophthalmic Plast Reconstr Surg 2014;30:246–250.

67 Baricich A, Cabrio C, Paggio R, Cisari C, Aluffi P: Peripheral facial nerve palsy: how effective is rehabilitation? Otol Neurotol 2012;33:1118–1126.

68 Lindsay RW, Robinson M, Hadlock TA: Comprehensive facial rehabilitation improves function in people with facial paralysis: a 5-year experience at the massachusetts eye and ear infirmary. Phys Ther 2010;90:391–397.

69 Volk GF, Finkensieper M, Guntinas-Lichius O: EMG biofeedback training at home for patient with chronic facial palsy and defective healing (in German). Laryngorhinootologie 2014;93:15–24.

70 Rosson GD, Redett RJ: Facial palsy: anatomy, etiology, grading, and surgical treatment. J Reconstr Microsurg 2008; 24:379–389.

Patrick J. Byrne, MD, FACS
Departments of Otolaryngology – Head and Neck Surgery and Dermatology
The Johns Hopkins University School of Medicine
601 N. Caroline St. 6th Floor
Baltimore, MD 21287 (USA)
E-Mail pbyrne2@jhmi.edu

Management of Regional Metastases of Malignant Salivary Gland Neoplasms

Jesus Medina[a] · Peter Zbären[b] · Patrick J. Bradley[c]

[a]Department of ORL-HNS, University of Oklahoma, Oklahoma City, Okla., USA; [b]Department of ORL-HNS, University Hospital, Bern, Switzerland; [c]Department of ORL-HNS, Nottingham University Hospitals, Nottingham, UK

Abstract

Metastases from salivary gland carcinomas to the cervical lymph nodes are relatively uncommon. However, their impact on prognosis is significant and, thus, it is important to manage them appropriately. Treatment of clinically evident metastases consists primarily of surgery, frequently followed by radiation. Management of the N0 neck, on the other hand, remains controversial. While there seems to be agreement regarding the tumor and patient factors that make it more likely for a patient to harbor subclinical metastases in the lymph nodes, some clinicians prefer to treat those patients with surgery, i.e. a neck dissection, and others prefer to use elective radiation. These different approaches and their rationale will be discussed in detail. © 2016 S. Karger AG, Basel

Introduction

Metastasis to the cervical lymph nodes is relatively uncommon in carcinomas of the salivary glands; around 15% of parotid cancers and 8–10% of submandibular and sublingual tumors present with clinical evidence of nodal metastasis [1]. Although rare, tumor involvement of the lymph nodes has a major influence on prognosis. A significant difference has been noted between the survival rates of patients with cancer of the parotid with and without histologically proven lymph node metastasis of 70 versus 10%, respectively. The corresponding rates for patients with submandibular gland cancer are 41 and 9%, respectively [2, 3]. Similar differences in the 5- and 10-year overall survival rates have been reported more recently [4, 5]. Therefore, appropriate management of the regional lymph nodes is important in the treatment of patients with carcinoma of the major salivary glands.

The Clinically Positive Neck

Treatment of salivary gland cancer with clinically or radiologically obvious lymph node metastases consists of neck dissection (ND), followed in most cases by postoperative radiation. The type of ND is determined by the extent of nodal disease, with the purpose of the operation being the removal of all gross tumor present. For patients with parotid carcinomas who have a single involved node in level II, dissection of levels II-IV may be appropriate, while dissection of levels I-V may be necessary in patients with multiple palpable nodes in different levels of the neck. In patients with submandibular gland carcinoma who have undergone therapeutic ND, lymph node metastases have been found in all neck levels. Level I is the most frequently involved; however, high rates of lymph node metastases in levels IV (40%) and V (25%) have been reported [6]. Complete resection of the involved nodes can often be accomplished with preservation of the spinal accessory nerve, the sternocleidomastoid muscle, or the internal jugular vein.

The addition of postoperative radiation appears to be of value for patients with salivary gland cancer who have cervical lymph node metastases. While there are no prospective or randomized studies to support this use of postoperative radiation, the results of several retrospective reports have indicated that it improves local-regional control and survival [7–9]. Armstrong et al. [10] performed matched-pair analysis of patients treated for salivary cancer with nodal metastasis, with one cohort of patients receiving surgery alone and the other cohort receiving postoperative radiation. Each cohort included 46 patients matched according to age and tumor type, grade and stage. Treatment of stage III and IV cancers with postoperative radiotherapy resulted in better loco-regional control and survival compared with treatment with surgery alone. However, patients with low-grade or early-stage (I/II) disease did not appear to benefit from the addition of radiation. Terhaard [11] analyzed the roles of primary and postoperative radiotherapy in 538 patients treated for salivary gland cancer in the Netherlands. The tumor was located in the parotid gland in 59% of the patients, the submandibular gland in 14%, the oral cavity in 23%, and elsewhere in 5%. In 386 of 498 patients, surgery was combined with radiotherapy, with a median dose of 62 Gy. Postoperative radiotherapy significantly improved regional control in the clinically positive (N+) neck (86 vs. 62% for surgery alone). A more recent study of 50 patients with parotid gland cancer treated in the United Kingdom demonstrated excellent local control (96%) with surgery and postoperative radiotherapy [12].

The beneficial effect of postoperative radiation has also been suggested in studies of patients with submandibular gland and minor salivary gland cancers [13–16]. It is unclear, however, whether radiation should be prescribed to every patient with histologically positive nodes [17] or only to those with multiple positive nodes or extranodal tumor extension [18].

Although the results of recent studies of the use of combined therapy for treatment of the N+ neck in salivary gland cancers look promising, they still leave a lot to be desired [10, 17]. In analysis of malignant tumors of major salivary gland origin, Armstrong et al. found a 5-year local-regional control rate of 69% for a group of 23 patients with lymph node metastases who received radiation to the neck and of 40% for a group of 16 patients treated with surgery alone (p = 0.05). The corresponding survival rates were 49 and 19%, respectively (p = 0.015). More recently, a single institution report has suggested that combining radiation with platinum-based chemotherapy may improve long-term survival among patients with locally advanced salivary gland carcinoma [19]. Thus, this treatment approach warrants investigation in patients with salivary gland carcinoma and clinically obvious lymph node metastases.

The Clinically Negative Neck

There is no general agreement to date about the management of the clinically negative (N0) neck in patients with cancer of the major salivary glands. The different management strategies that have been recommended over the years are outlined in table 1.

In a paper published in 1967 reporting a study that included 111 patients with malignant tumors of the parotid gland treated at MD Anderson Cancer Center, Bardwil [20] stated that elective ND had not been performed during the last 5 years of the study because he had found occult metastases in only 1 of 34 (2.9%) patients who had undergone this operation. Interestingly, however, he advocated 'careful dissection of the first echelon of lymph nodes for all lesions,' claiming that this adds little or no morbidity to the operation and if the tumor proves to be malignant, the procedure is generally adequate. For the next 2 decades, the surgeons at MD Anderson subscribed to this approach to the N0 neck in patients with salivary gland tumors [18, 21]. Then, in 1993, Frankenthaler et al. [22] reported the results of multivariate analysis of 11 clinical and histopathologic variables in patients who had undergone elective node dissection for cancer of the parotid gland. The factors that were correlated with the presence of occult cervical lymph node metastases were facial paralysis, an older age (>54 years), a high tumor grade, perilymphatic invasion, and extraparotid tumor extension. Interestingly, with the exception of older age, these factors were also associated with increased local recurrence and thus dictate the need for postoperative radiation, regardless of the presence or absence of occult metastases in the regional lymph nodes [23]. Another interesting finding of this study was that occult metastases were discovered by elective dissection of the lymph nodes in only 3% of the patients with low-grade tumors. Although staging ND may be helpful for determining the need for postoperative radiation in this group of patients, the operation would be unnecessary in 97% of the cases. In 1980, Johns [24] carried out a review of the literature and of the experience of the University of Virginia with the treatment of parotid tumors. He advocated treating the neck on the basis of the stage and histology of the primary tumor and felt that ND was indicated in patients with tumors of stages T3 and T4; for T1 through T2 tumors of high-grade histology, he advocated dissection of the first echelon of lymph nodes [24]. In 1989, Spiro et al. [25] published a review of 44 years of experience at the Memorial Sloan-Kettering Cancer Center. This group stated that elective ND may be beneficial in patients with anaplastic or squamous carcinoma because 58% or more of them will develop cervical metastases. Staging supraomohyoid ND was thought to be appropriate for patients with other types of high-grade tumors [25]. A few years later, similar recommendations were made by Califano et al. [26] from the University of Naples. A later study from Memorial Sloan-Kettering suggested that in patients with small, low-grade tumors that are adequately excised, elective treatment of the neck is not necessary [23]. On the other hand, this treatment is warranted in patients whose tumors are larger than 4 cm or are high grade and in whom the risk of occult lymph node metastases is high [24]. In yet a different review of the same patient population, Armstrong et al. [1] advanced the notion that in the high-risk group, the N0 neck could be treated with either elective ND (incorporating at least levels I, II, and III) or 'elective postoperative neck irradiation.' They suggested that it may be reasonable to treat the neck electively with radiation for patients for whom postoperative radiation therapy is indicated according to the characteristics of the primary tumor. Their recommendation is based on the extensive experience reported, which indicates that either radiation therapy to the neck or ND can control clinically occult cervical metastases in epidermoid carcinoma of the head and neck.

Table 1. The N0 neck in salivary gland cancer: management strategies over time

Bardwill [20], 1967	No elective radical neck dissection Dissection of first echelon nodes for all tumors
Johns [24], 1980	T1–T2: No neck dissection T3–T4: Neck dissection and radiation
Byers [21], 1982	Dissection of first echelon nodes in all tumors
Spiro et al. [25], 1989	Elective neck dissection for anaplastic or squamous carcinoma Staging supraomohyoid neck dissection for other high-grade tumors
Armstrong et al. [1], 1992	Elective neck dissection or elective postoperative neck irradiation for 'high-risk' tumors
Califano et al. [26], 1993	Neck dissection in cases of mucoepidermoid, anaplastic, and squamous cell carcinomas
Ball et al. [41], 1995	Neck dissection: high-grade tumors with positive jugulodigastric node biopsy (intraoperative frozen section)
Frankenthaler et al. [22], 1993	No elective neck dissection Elective postoperative neck irradiation for high-risk tumors
Kelley and Spiro [17], 1996	Elective treatment of the neck for tumors larger than 4 cm or high-grade tumors
Medina [33], 1998	Intraoperative assessment of level II nodes If suspicious and frozen section examination reveals metastases: ND Otherwise, elective postoperative neck irradiation if the primary tumor exhibits high-risk clinical-pathological characteristics
Wang et al. [42], 2012	Comprehensive ND and postoperative radiation: high-risk tumors with adverse features Upper ND; postoperative radiation: moderate-risk tumors, depending on adverse factors and pN status Observation: low-risk tumors without adverse features
Herman et al. [27], 2013	T3–T4: Elective neck irradiation if postoperative radiation is indicated preoperatively (based on primary tumor characteristics)
Norbis et al. [28], 2014	Elective neck dissection for all patients

As shown in table 1, the debate continues to date, with some clinical studies suggesting that the appropriate treatment of the N0 neck is either ND or postoperative neck irradiation in select cases [27] and others suggesting that elective ND should be performed in all cases [28].

An important consideration in selective approaches to management of the N0 neck is the definition of a 'high-risk' tumor with regard to the risk of occult metastases in the lymph nodes. Today, it is generally accepted that the tumor histology, T stage and tumor grade are the most consistent predictors of the presence of nodal metastases. The risk of metastases is approximately 50% or higher when the tumor histology is undifferentiated carcinoma, adenocarcinoma, salivary duct carcinoma or squamous cell carcinoma. The risk is similarly high for high-grade mucoepidermoid carcinoma, unlike low-grade mucoepidermoid carcinoma and acinic cell car-

Table 2. Variables correlated with the presence of lymph node metastases in cancers of the major salivary glands

1	High-grade tumors
2	T3 (?) and T4 tumors
3	Facial paralysis
4	Older age (>54 years [22], >70 years [43])
5	Extraparotid extension [42]
6	Lymphovascular invasion
7	Facial paralysis and perineural invasion [22, 33]

cinoma, for which the risk is 2–4% [1, 22, 29]. The reported prevalence of lymph node metastases also varies according to the disease stage, ranging from 16 to 33% for T3 tumors and from 24 to 50% for T4 tumors [1, 8, 26]. Beppu et al. [30] found that the incidence of lymph node metastases in patients with submandibular gland carcinoma ranged from 0% in T1 tumors to 33.3% in T2, 57.1% in T3 and 100% in T4 tumors. Other factors that have been reported to be correlated with the presence of lymph node metastases in salivary gland tumors are listed in table 2.

It is interesting to note that the currently accepted indications for prescribing postoperative radiation in patients with carcinoma of the salivary gland also include high risk and high-grade histology, stages T3 and T4, extraparotid extension, perineural invasion, a deep lobe location, close or positive margins and the presence of lymph node metastases [31–33].

Consequently, Medina [33] reasoned, along the lines suggested by Armstrong et al. [1], that since the characteristics of salivary gland carcinomas that dictate the need for elective treatment of the regional lymph nodes are, in essence, the same characteristics that dictate the need for postoperative radiation to the primary lesion, then it seems reasonable to treat the neck with elective irradiation in patients whose tumors exhibit these characteristics after adequate surgery of the primary tumor. Although this approach to the N0 neck seems logical, it has been criticized because it has not been tested prospectively, and it is assumed that elective neck irradiation is efficacious in the control of occult lymph node metastases from salivary gland carcinomas. Interestingly, two recent studies, albeit retrospective, have lent support to this approach. Chen et al. [31] studied 251 patients with carcinoma of the salivary glands and a clinically N0 neck who were treated with surgery and postoperative radiation therapy and who had not undergone previous ND. Postoperative elective neck irradiation reduced the 10-year nodal failure rate from 26 to 0% (p = 0.0001). The highest crude rates of nodal relapse among patients treated without elective neck irradiation were observed in those with squamous cell carcinoma (67%), undifferentiated carcinoma (50%), adenocarcinoma (34%), and mucoepidermoid carcinoma (29%). There were no nodal failures observed among patients with adenoid cystic or acinic cell carcinoma. These authors concluded that elective postoperative irradiation effectively prevents nodal relapses and should be used for select patients at high risk of regional failure. The second study was conducted by Herman et al. [27] to determine whether patients with high-grade salivary gland carcinoma and a clinically node-negative neck benefit from elective ND performed prior to postoperative radiotherapy. They studied 59 previously untreated patients with high-grade salivary gland carcinoma and an N0 neck who were treated with curative intent using elective ND (n = 41) or elective neck irradiation (n = 18). These patients underwent resection of the primary tumor followed by postoperative radiation. During a median follow-up period of 5.2 years (range, 0.3–34 years), there were 4 recurrences (10%) in the ND group and 0 in the neck irradiation group. They concluded that patients with high-grade salivary gland carcinoma and a clinically N0 neck who have undergone surgery and postoperative radiation are not likely to benefit from ND [27].

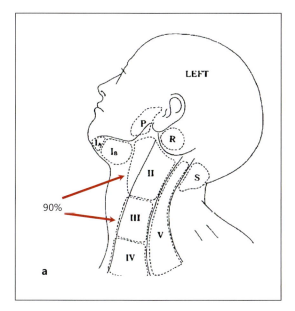
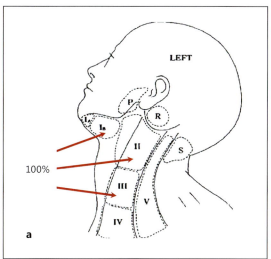
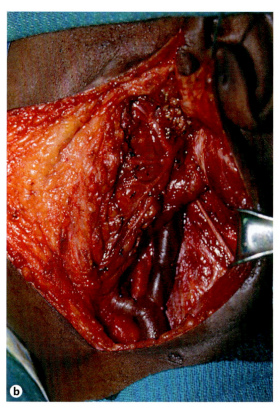
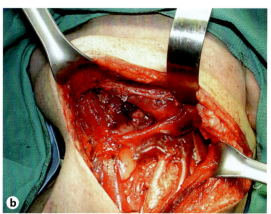

Fig. 1. a Distribution of occult lymph node metastases in parotid cancers. **b** Intraoperative photograph of parotidectomy and selective neck dissection of levels II and III.

Fig. 2. a Distribution of occult lymph node metastases in carcinomas of the submandibular gland in levels II and III. **b** Intraoperative photograph of selective neck dissection of levels I (submandibular triangle), II and III.

Sentinel Node Biopsy in Primary Parotid Cancer

The incidence of occult metastasis in elective neck dissection (END) specimens has been reported to be between 20 and 37% [34, 35]. Thus, the performing of END for all cN0 parotid cancers implies that more than half of patients are over-treated. Sentinel node biopsy (SNB) is a

Salivary Gland: Neck

technique used to select patients with occult metastases who would benefit from ND. The sentinel lymph node is the first node (in many cases, there are several sentinel nodes) to receive drainage from the primary tumor site. It is therefore the initial possible recipient of metastatic tumor cells and may be predictive of the histopathologic status of the remaining lymphatic neck area [36]. The SNB concept was introduced for nodal evaluation of head and neck squamous cell carcinomas localized to the oral cavity and oropharynx. It was determined to be a valid procedure for avoiding END and its potential morbidity [36]. As for primary parotid carcinoma, experience with SNB is very limited, although one of the first such reported experiences involved patients treated for parotid carcinoma [37]. Gould et al. [37] observed the so-called angular node in 28 parotid carcinomas; they based the decision on whether to perform ND on the results of frozen section analysis of this node. Thereafter, several authors suggested 'dissection of the first echelon of lymph nodes' [20, 24] or 'jugulodigastric node biopsy' with intraoperative frozen section analysis [33].

Approximately 80% of parotid carcinomas arise in the superficial lobe. In many of these cases, the first echelon is represented by the intraparotid lymph nodes. The mean numbers of intraparotid lymph nodes that have been reported are 7 (range, 3–19) in the superficial lobe and 2 (range, 0–9) in the deep lobe [38]. Furthermore, a statistically significant correlation ($p = 0.005$) has been observed between the presence of intraparotid metastatic nodes and neck node metastases [39]. Unfortunately, the lymphoscintigraphic identification of intraparotid nod metastases is challenging in many cases. The nodes can be easily missed due to close proximity to the primary tumor and injection site.

In 2006, Starek et al. [40] reported a study of 6 patients who underwent SNB; in all cases, selective neck dissection level II–V was performed. Two patients had true-positive results for the sentinel node, and there was one false-negative result that was interpreted as 'distortion of the lymphatic outflow resulting from intraparotid localization of lymphatic metastases'.

There are several documented reasons why there is no experience with performing SNB for parotid carcinoma: the lymphoscintigraphic identification of intraparotid lymph nodes may be difficult in many cases; intraparotid lymph nodes can be examined by frozen section analysis of a parotidectomy specimen; and lymphadenectomy of suspected nodes in level II for frozen sections or even END of level II or level III can easily be incorporated into the surgical approach with minimal morbidity and only a slight increase in operative time.

Conclusion

Regardless of whether a clinician chooses to treat an N0 neck with elective ND or elective postoperative radiation, only the ipsilateral side of the neck should be treated. The occurrence of contralateral lymph node metastases in tumors of the major salivary gland is negligible [17].

If a clinician elects to treat an N0 neck with ND, the type of ND should be tailored according to the distribution of occult metastases in patients with salivary gland carcinoma. The distribution of lymph node metastases in parotid carcinoma has been studied by Armstrong et al. [1]. Beppu et al. [30] noted that pathological neck lymph node metastases were found in levels II and III only in 27 patients with submandibular gland carcinoma staged as N0. Others have reported similar findings [6] (fig. 1, 2).

References

1 Armstrong JG, Harrison LB, Thaler HT, et al: The indications for elective treatment of the neck in cancer of the major salivary glands. Cancer 1992;69:615–619.
2 Shah J: Management of regional metastasis in salivary and thyroid cancer; in: Larson DA, Ballantyne AJ, Guillamondegui OM (eds): Cancer in the Neck: Evaluation and Treatment. New York, Macmillan, 1986, pp 253–258.
3 Larson D, Ballantyne A, Guillamondegui O: Cancer in the Neck: Evaluation and Treatment. New York, Macmillan, 1986, pp 253–258.
4 Korkmaz H, Yoo GH, Du W, et al: Predictors of nodal metastasis in salivary gland cancer. J Surg Oncol 2002;80:186–189.
5 Stennert E, Kisner D, Jungehuelsing M, et al: HIgh incidence of lymph node metastasis in major salivary gland cancer. Arch Otolaryngol Head Neck Surg 2003;129:720–723.
6 Han MW, Cho KJ, Roh JL, et al: Patterns of lymph node metastasis and their influence on outcomes in patients with submandibular gland carcinoma. J Surg Oncol 2012;106:475–480.
7 Byers RM: Symposium: adjuvant cancer therapy of head and neck tumors. The use of postoperative irradiation – its goals and 1978 attainments. Laryngoscope 1979;89:567–572.
8 Fu KK, Leibel SA, Levine ML, et al: Carcinoma of the major and minor salivary glands: analysis of treatment results and sites and causes of failures. Cancer 1977;40:2882–2890.
9 King JJ, Fletcher GH: Malignant tumors of the major salivary glands. Radiology 1971;100:381–384.
10 Armstrong J, Harrison L, Spiro R, et al: Malignant tumors of major salivary gland origin. Arch Otolaryngol Head and Neck Surg 1990;116:290–293.
11 Terhaard CHJ: Postoperative and primary radiotherapy for salivary gland carcinomas: indications, techniques and results. Int J Radiat Oncol Biol Phys 2007;89(2 suppl):S52–S55.
12 Shah K, Javed F, Alcock C, et al: Parotid cancer treatment with surgery followed by radiotherapy in Oxford over 15 years. Ann R Coll Surg Engl 2011;93:218–222.
13 Bissett RJ, Fitzpatrick PJ: Malignant submandibular gland tumors: a review of 91 patients. Am J Clin Oncol 1988;11:46–51.
14 Zeidan YH, Shultz DB, Murphy JD, et al: Long-term outcomes of surgery followed by radiation therapy for minor salivary gland carcinomas. Laryngoscope 2013;123:2675–2680.
15 Garden AS, Weber RS, Ang KK, et al: Postoperative radiation therapy for malignant tumors of minor salivary glands. Outcome and patterns of failure. Cancer 1994;73:2563–2569.
16 Parsons JT, Mendenhall WM, Stringer SP, et al: Management of minor salivary gland carcinomas. Int J Radiat Oncol Biol Phys 1996;35:443–454.
17 Kelley DJ, Spiro RH: Management of the neck in parotid carcinoma. Am J Surg 1996;172:695–697.
18 Jackson GL, Luna MA, Byers RM: Results of surgery alone and surgery combined with postoperative radiotherapy in the treatment of cancer of the parotid gland. Am J Surg 1983;146:497–500.
19 Tanvetyanon T, Qin D, Padhya T, et al: Outcomes of postoperative concurrent chemoradiotherapy for locally advanced major salivary gland carcinoma. Arch Otolaryngol Head Neck Surg 2009;135:687–692.
20 Bardwil JM: Tumors of the parotid gland. Am J Surg 1967;114:498–502.
21 Byers R: Treatment of malignant tumors of the parotid and submaxillary glands. Resident Staff Physician 1982;28:52.
22 Frankenthaler RA, Byers RM, Luna MA, et al: Predicting occult lymph node metastasis in parotid cancer. Arch Otolaryngol Head Neck Surg 1993;119:517–520.
23 Frankenthaler RA, Luna MA, Lee SS, et al: Prognostic variables in parotid cancer. Arch Otolaryngol Head Neck Surg 1991;117:1251–1256.
24 Johns ME: Parotid cancer: a rational basis for treatment. Head Neck Surg 1980;3:132–141.
25 Spiro RH, Armstrong J, Harrison L, et al: Carcinoma of major salivary glands. Recent trends. Arch Otolaryngol Head Neck Surg 1989;115:316–321.
26 Califano L, Zupi A, Massari PS, et al: Indication for neck dissection in carcinoma of the parotid gland. Our experience on 39 cases. Int Surg 1993;78:347–349.
27 Herman MP, Werning JW, Morris CG, et al: Elective neck management for high-grade salivary gland carcinoma. Am J Otolaryngol 2013;34:205–208.
28 Norbis C-P, Rohleder NH, Wolff D, et al: Head and neck salivary gland carcinoma – elective neck dissection, yes or no? J Oral Maxillofac Surg 2014;72:205–210.
29 Regis de Brito Santos I, Kowalski LP, Cavalcante de Araujo V, et al: Multivariate analysis of risk factors for neck metastases in surgically treated parotid carcinoma. Arch Otolaryngol Head Neck Surg 2001;127:56–60.
30 Beppu T, Kamata SE, Kawabata K, et al: Prophylactic neck dissection for submandibular gland cancer (in Japanese). Nihon Jibiinkoka Gakkai Kaiho 2003;106:831–837.
31 Chen AM, Granchi PJ, Garcia J, et al: Local-regional recurrence after surgery without postoperative irradiation for carcinomas of the major salivary glands: Implications for adjuvant therapy. Int J Radiat Oncol Biol Phys 2007;67:982–987.
32 Ferlito A, Pellitteri PK, Robbins KT, et al: Management of the neck in cancer of the major salivary glands, thyroid and parathyroid glands. Acta Otolaryngol 2002;122:673–678.
33 Medina JE: Neck dissection in the treatment of cancer of the major salivary glands. Otolaryngol Clin North Am 1998;31:815–822.
34 Klussmann JP, Ponert T, Muller RP, et al: Patterns of lymph node spread and its influence on outcome in resectable parotid cancer. Eur J Surg Oncol 2008;34:932–937.
35 Zbaren P, Schupbach J, Nuyens M, et al: Elective neck dissection versus observation in primary parotid carcinoma. Otolaryngol Head Neck Surg 2005;132:387–391.
36 Stoeckli SJ: Sentinel lymph node biopsy for oral and oropharyngeal squamous cell carcinoma of the head and neck. Laryngoscope 2007;117:1539–1551.

37 Gould EA, Winship T, Philbin PH, et al: Observations on a 'sentinel node' in cancer of the parotid'. Cancer 1960;13:77–78.
38 Olsen KD, Moore EJ: Deep lobe parotidectomy: clinical rationale in the management of the primary and metastatis cancer. Eur Arch Otorhinolaryngol 2014;271:1181–1185.
39 Lim CM, Gilbert MR, Johnson JT, et al: Clinical significance of intraparotid lymph node metastasis in primary parotid cancer. Head Neck 2014;36:1634–1637.
40 Hornstra MT, Alkureishi LW, Ross GL, et al: Predictive factors for failure to identify sentinel nodes in head and neck squamous cell carcinoma. Head Neck 2008;30:858–862.
41 Ball AB, Fish S, Thomas JM: Malignant epithelial parotid tumours: a rational treatment policy. Br J Surg 1995;82:621–623.
42 Wang Y-L, Li D-S, Gan H-L, et al: Predictive index for lymph node management of major salivary gland cancer. Laryngoscope 2012;122:1497–1506.
43 Ettl T, Gosau M, Brockhoff G, et al: Predictors of cervical lymph node metastasis in salivary gland cancer. Head Neck 2014;36:517–523.

Jesus Medina
Department of ORL-HNS
University of Oklahoma, 920 Stanton L. Young Blvd, WP1290
Oklahoma City, OK 73104 (USA)
E-Mail Jesus-medina@ouhsc.edu

Indications for Salivary Gland Radiotherapy

David J. Thomson[a, b] · Nick J. Slevin[a, b] · William M. Mendenhall[c]

[a]Department of Clinical Oncology, The Christie NHS Foundation Trust, Manchester, and [b]Institute of Cancer Sciences, University of Manchester, Manchester, UK; [c]Department of Radiation Oncology, College of Medicine, University of Florida, Gainesville, Fla., USA

Abstract

There is an established role for post-operative radiotherapy in the treatment of benign and malignant salivary gland tumours. For benign disease, the addition of radiotherapy improves local tumour control in cases with incomplete excision, involved surgical margins or multifocal disease recurrence. After capsule rupture or spillage alone, surveillance should usually be advised. For malignant disease, post-operative radiotherapy is recommended for an advanced tumour stage, high-grade tumour, perineural or lympho-vascular invasion, close or positive resection margins, extra-parotid extension or lymph node involvement. The main benefit is increased loco-regional tumour control, although this may translate into a modest improvement in survival. The possible late side effects of parotid bed irradiation include skin changes, chronic otitis externa, sensorineural hearing loss, osteoradionecrosis and secondary malignancy. Severe complications are rare, but patients should be counselled carefully about the risks. Primary radiotherapy is unlikely to be curative and is reserved to cases in which resection would cause unacceptable functional or cosmetic morbidity or would likely result in subtotal resection (R2) or to patients with distant metastases to gain local tumour control. There are provisional data on the use of charged particle radiotherapy in this setting. Some patients may benefit from synchronous chemotherapy with radiotherapy, but this group is not defined, and data from comparative prospective studies are required before routine clinical use of this treatment.

© 2016 S. Karger AG, Basel

Radiotherapy for Benign Disease

Parotid gland tumours account for approximately 70% of all salivary gland tumours, 70% of which are benign, and 85% of these benign tumours are pleomorphic adenomas [1]. The definitive treatment is surgery, which aims to achieve a histological diagnosis, local tumour control and prevent malignant transformation, which is seen in approximately 3–4% of cases [2].

Post-Operative Radiotherapy

Defining the role of post-operative radiotherapy is challenging due to small patient numbers and difficulties with interpretation of data from retrospective series. Complete microscopic resection achieves local tumour control in over 95% of cas-

Table 1. Post-operative radiotherapy for pleomorphic adenoma

For	Against
Decreased rate of loco-regional recurrence (to less than 5%)	Long natural history favours watch and wait policy
Reduced chance of repeat surgery and damage to VII nerve	Possible late effects of radiotherapy, particularly sensorineural hearing loss
Reduced rate of malignant transformation, seen in 3–4% of cases	Risk of radiation-induced malignancy, seen in 1% of cases

es [3–5]. Post-operative radiotherapy is not advised in this setting because the absolute local control benefit is small, and there is potential for late treatment-related morbidity.

In the few patients with incompletely excised tumours or involved surgical margins, post-operative radiotherapy is recommended (table 1). Long-term follow-up is required because there is a propensity for late local recurrence. In an Edinburgh series of 311 patients with parotid pleomorphic adenoma treated with local excision and radiotherapy, there were 10 benign recurrences. These recurrences occurred in approximately 1% of at-risk patients within each 5-year follow-up period of 0–5, 5–10, 10–15 and 15–20 years [6].

In a study of 115 patients who received post-operative radiotherapy after incomplete tumour removal or spillage, the local recurrence rate was 0.9%, with a median follow-up time of 14 years [7]. For 14 cases with involved margins or tumour spillage, the addition of post-operative radiotherapy was associated with a reduction in the rate of local tumour recurrence (13 vs. 0%, mean follow-up time of 7.4 years) [8]. In a more recent series, the use of post-operative radiotherapy to treat 25 out of 78 patients with incomplete tumour excision (83%) or capsule rupture alone (17%) also reduced the rate of local tumour recurrence (21% vs. 4%, median follow-up time of 6.4 years) [9].

Isolated tumour capsule rupture or spillage only moderately increases the incidence of local tumour recurrence. In this setting, surveillance following surgery may be advised. In a series from Buchman et al. [8], there were no local recurrences in a subset of four patients with tumour spillage treated with surgery without post-operative radiotherapy. In a larger series with an overall mean follow-up time of 18 years, among 26 out of 238 patients with capsule rupture and macroscopic tumour spillage, two (8%) developed recurrent disease [10]. This rate was greater but was not significantly different from the overall local recurrence rate of 2%. The overall time to recurrence from surgery was between 7 and 18 years. In matched pairs analysis of 60 cases treated with total parotidectomy, partial superficial parotidectomy or extracapsular dissection, capsule rupture or tumour spillage occurred in 2 of 20, 1 of 20 and 1 of 20 cases, respectively, and after a median follow-up time of 8.3 years, none of the cases demonstrated local recurrence [4]. In a 20-year review of the literature (1979–1999) and analysis of 4,608 cases, tumour spillage was associated with only a modestly increased rate of local recurrence (5 vs. 2–3%, excluding patients treated with enucleation) [4]. It is also of note that capsule exposure at the facial nerve interface is common, independent of the surgical approach used or extent of parotid gland sacrifice [4]. This factor alone does not warrant the addition of post-operative radiotherapy.

Radiotherapy for Post-Operative Recurrent Pleomorphic Adenoma

For recurrent disease, a second surgery is generally indicated to gain local tumour control and due to the apparent increased risk of malignant transformation in this sub-group, as seen in

0–23% of cases [11]. Post-operative radiotherapy is advised for multi-nodular recurrence to improve local tumour control. In a retrospective series including 33 patients with recurrent pleomorphic adenoma, 16 received post-operative radiotherapy, 13 (81%) of whom achieved local tumour control, which was attained in only 1 (6%) patient treated with surgery alone [12]. Similarly, in a retrospective series of 31 patients, local disease control was seen in all 11 patients (100% at 10 years) who received post-operative radiotherapy but in only 14 out of 20 patients (71% at 10 years) treated with surgery alone [13]. Renehan et al. [14] reported a reduction in the local recurrence rate with the addition of post-operative radiotherapy for 114 patients (24 vs. 8% at 15 years). However, the benefit was only seen for multi-nodular and not unifocal recurrence. Surveillance after surgery may also be appropriate for elderly patients or those with extensive co-morbidities.

When complete macroscopic resection is not possible or would require facial nerve sacrifice, primary radiotherapy may be used [11]. However, in the presence of macroscopic disease, it is unlikely to be curative. Following surgery and post-operative radiotherapy for recurrent disease, 16 out of 17 (94%) patients with residual microscopic disease have been reported to remain disease free, whereas only 1 out of 4 (25%) patients with gross residual disease have been shown to achieve local disease control [15]. Post-operative radiotherapy is also recommended after multiple recurrences to reduce the chance of further relapse because subsequent surgery is less effective and is associated with an increased risk of permanent facial nerve damage [16].

Radiotherapy for Malignant Disease

Salivary gland malignancy is rare and accounts for approximately 5% of all head and neck cancers [17]. Most cases occur in the parotid gland, with the remainder occurring in the submandibular, sublingual and minor salivary glands.

Radiotherapy is mainly used in the post-operative setting to treat possible microscopic residual disease and improve loco-regional control [18–25]. For advanced stage and/or high-grade disease, this may translate into improved survival [22–24]. The use of primary radiotherapy is reserved for unresectable disease, some minor salivary gland cancers for which resection would result unacceptable functional or cosmetic morbidity, and palliation [25, 26].

The evidence base is limited by small patient numbers, tumour and treatment heterogeneity and retrospective analyses. The indications for post-operative radiotherapy include the following: advanced stage (T 3–4), high grade, perineural or lympho-vascular invasion, close (less than 5 mm) or positive resection margins and extra-parotid extension. Adjuvant neck radiotherapy should be considered in the presence of one or more pathologically involved lymph nodes [25].

Post-Operative Radiotherapy for Salivary Gland Cancer

Matched pair analysis was carried out on 46 patients who received surgery and post-operative radiotherapy for major salivary gland cancer at the Memorial Sloan Kettering Cancer Centre [22]. The patients were matched by prognostic criteria with patients treated with surgery alone. The median follow-up times were 10.5 years (surgery alone) and 5.8 years (post-operative radiotherapy). For stages III and IV disease only, the 5-year local control rate was improved by addition of radiotherapy (51 vs. 17%). This translated into improved 5-year disease-specific survival (51 vs. 10%), with a trend towards benefiting patients with high-grade disease (57 vs. 28%).

In a retrospective analysis of 35 patients with major or minor salivary gland cancer with close or involved surgical margins treated at the University of California at San Francisco, post-operative radiotherapy significantly reduced the rate of local recurrence (14 vs. 54%) [18]. In a subsequent series (1960–1977), 38 patients with early-stage major salivary gland cancer underwent surgery alone, and 30 patients with high-grade or advanced-stage disease received post-operative radiotherapy [19]. After a minimum follow-up time of 2 years, the respective proportions of patients who were alive and disease-free were 57 and 70%. In a retrospective series of 120 patients who underwent surgery for parotid cancer, the addition of post-operative radiotherapy in 59 (49%) cases was associated with improved local control in those with locally advanced or high-grade disease [20]. In a review of 103 patients with parotid cancer treated surgically at The Christie from 1952 to 1992, two-thirds of the patients received post-operative radiotherapy [24]. With a median follow-up time of 12 years, the addition of post-operative radiotherapy significantly reduced loco-regional tumour recurrence (15 vs. 43%). All of the patients benefited from this treatment, but the effect was the most pronounced for the patients with tumours that were greater than 4 cm in size, high-grade disease or adenoid cystic carcinoma.

A retrospective study of patients with malignant salivary gland cancer who were treated with surgery alone (67 patients) or with post-operative radiotherapy (169 patients) at the Princess Margaret Hospital between 1958 and 1980 demonstrated, with a median follow-up time of 10 years, an improved 10-year relapse-free rate in the patients who received radiotherapy (62 vs. 22%) [21]. While there was possible selection bias and incomplete staging, the known prognostic factors were balanced between the groups. Similarly, in a subsequent series from Johns Hopkins Hospital, of 69 patients with previously untreated major salivary gland cancer, 5 of 19 (26%) treated with surgery alone had local recurrence, whereas only 2 of 50 (4%) had local recurrence following post-operative radiotherapy [23]. These findings were associated with an improvement in the 5-year disease-specific actuarial survival (75 vs. 59%). In a large retrospective series from the Netherlands, patients with major or minor salivary gland cancer were treated with surgery alone (112 patients, 22%) or with both surgery and post-operative radiotherapy (386 patients, 78%) [25]. The prognostic factors favoured the patients who received surgery alone. However, the actuarial local control rates after 5 and 10 years were 84 and 76%, respectively, for surgery alone and were 94 and 91%, respectively, with the addition of post-operative radiotherapy. Multivariate analysis revealed that post-operative radiotherapy significantly improved 10-year local control for T3–4 tumours, close (less than 5 mm) or involved margins, and perineural or bone invasion; further, regional control was significantly improved in the presence of one or more involved neck nodes.

Single institution outcomes for patients with salivary gland cancer treated with surgery and post-operative radiotherapy demonstrate an important role for radiotherapy, with approximate overall local tumour control of 85–90% (table 2) [26–29]. Advanced tumour stage was an adverse prognostic factor.

Late Toxicities of Parotid Bed Radiotherapy
The possible late side effects of parotid bed irradiation include skin changes, chronic otitis externa and sensorineural hearing loss. These effects are mostly subclinical, but in pooled analysis, 36% of patients developed hearing loss of 10 dB or more at 4 kHz [30]. Osteoradionecrosis of the jaw is unusual, and in one series, it was observed in 2 of 106 patients with malignant salivary gland tumours treated initially with parotidectomy [31]. The rate of radiotherapy-induced malignancy was approximately 1% at long-term follow-up. This finding is particularly important, and this relatively young patient population should be counselled carefully for the management of benign disease. The Christie reported that among 115 patients treated with surgery and post-operative radiotherapy, over a median follow-up time of 14 years, one patient (1%) developed squamous

Table 2. Local control of salivary gland cancer with surgery and post-operative radiotherapy

Institution and reference	Salivary gland site	Patients, n	Local control at 10 years (median follow-up)
MD Anderson [28]	Major	166	90% (12.9 years)
The Netherlands [25]	Major or minor	386	91% (mean, 8.8 years)
University of Florida [26]	Major or minor	160	90% (8.3 years)
Stanford University [29]	Minor	90	88% (5.9 years)
Memorial Sloan-Kettering [27]	Minor	98	83% (7.3 years)

cell carcinoma of the mandible within the irradiated area [7]. Similarly, in a series from Edinburgh, there were four (1%) cases of malignant recurrence at 14–18 years after treatment [6].

Primary Radiotherapy for Salivary Gland Cancer
Primary radiotherapy is reserved for patients for whom resection would result in unacceptable functional or cosmetic morbidity or those who are unfit for surgery or to gain local tumour control in those with distant metastases. In a series from the University of Florida, among 64 patients treated with radiotherapy alone, the 10-year loco-regional control rate was 40% [26]. Regression analysis revealed that this rate was highly dependent on the stage of disease. The 10-year loco-regional control rates for stage I–III and IV disease were 70 and 24%, respectively. Terhaard et al. [25] correlated the local disease responses at 3–6 weeks after treatment with the 5-year local control rate. Among 40 patients, 38% demonstrated a complete response, and the 5-year actuarial local control rate was 69% for this group. Among the remainder of the patients, local control was observed in only 10%. The overall 5-year local control rate was approximately 30%.

Charged Particle Therapy
The low proliferation rate and increased proportion of salivary gland cancer cells during the resting state of the cell cycle makes the use of high linear energy transfer neutron or carbon ion radiation attractive. The Radiation Therapy Oncology Group and Medical Research Council phase III trial compared fast neutron therapy with photon irradiation in unresectable salivary gland cancer [32]. For 25 patients over a minimum follow-up time of 6.3 years, the actuarial 10-year loco-regional control rate was superior for the neutron group (56 vs. 17%), but there was no difference in overall survival (15 vs. 25%). The rates of severe toxicities, including mucosal reactions and fibrosis, were greater with neutrons. The apparent absence of a survival benefit, increased morbidity and lack of treatment facilities prevented the widespread adoption of this therapy.

In contrast with neutrons, for carbon ions, the linear energy transfer is not uniformly high but builds to a maximum at the Bragg peak. This is dosimetrically advantageous and has the potential to reduce treatment-related toxicities. The use of carbon ions was investigated in patients with locally advanced adenoid cystic carcinoma (all but three had tumours infiltrating the skull base, and in these patients, the disease infiltrated the orbits) [33]. As part of a phase I–II trial, 29 patients received combined stereotactic fractionated photon radiotherapy with concomitant carbon ion boost. The outcomes were compared with those of 34 patients who received stereotactic fractionated (intensity-modulated) radiotherapy alone. For both groups, the treatments were well tolerated, and the rates of severe late toxicities were less than 5%. There was a trend of improvement in loco-regional control in the carbon ion group (78 vs. 72 and 78 vs. 25% at 2 and 4 years, respectively). However, whether this improvement was due to an enhanced biological effect of the carbon ions or increased precision and conformity, enabling dose intensification to the gross tumour volume (72 GyE vs. 66 Gy), was uncertain.

The advantageous dose depth distribution and precision afforded by proton beam therapy makes it an attractive modality for salivary gland cancer when in proximity to the dose-limiting visual apparatus. In a retrospective study from the Massachusetts General Hospital, 23 patients with adenoid cystic carcinoma involving the skull base received combined photon and proton beam radiotherapy [34]. The median achieved total dose to the primary site was 75.9 GyE, and the 5-year local control rate was 93%.

Addition of Chemotherapy

Studies have generally failed to demonstrate a benefit of the addition of chemotherapy to either definitive or adjuvant radiotherapy [35]. However, the data are limited to small retrospective or non-randomized phase II studies and are subject to selection bias. Some patients may benefit from this treatment, and the strongest basis for chemotherapy favours cisplatin. However, this patient group is not defined, and data from prospective comparative studies are required before routine clinical use of this treatment.

References

1 Spiro RH: Salivary neoplasms: overview of a 35-year experience with 2,807 patients. Head Neck Surg 1986;8:177–184.
2 Seifert G: Histopathology of malignant salivary gland tumours. Eur J Cancer B Oral Oncol 1992;28B:49–56.
3 Renehan A, Gleave EN, Hancock BD, et al: Long-term follow-up of over 1,000 patients with salivary gland tumours treated in a single centre. Br J Surg 1996; 83:1750–1754.
4 Witt RL: The significance of the margin in parotid surgery for pleomorphic adenoma. Laryngoscope 2002;112:2141–2154.
5 O'Brien CJ: Current management of benign parotid tumors – the role of limited superficial parotidectomy. Head Neck 2003;25:946–952.
6 Dawson AK, Orr JA: Long-term results of local excision and radiotherapy in pleomorphic adenoma of the parotid. Int J Radiat Oncol Biol Phys 1985;11: 451–455.
7 Barton J, Slevin NJ, Gleave EN: Radiotherapy for pleomorphic adenoma of the parotid gland. Int J Radiat Oncol Biol Phys 1992;22:925–928.
8 Buchman C, Stringer SP, Mendenhall WM, et al: Pleomorphic adenoma: effect of tumor spill and inadequate resection on tumor recurrence. Laryngoscope 1994;104:1231–1234.
9 Robertson BF, Robertson GA, Shoaib T, et al: Pleomorphic adenomas: post-operative radiotherapy is unnecessary following primary incomplete excision: a retrospective review. J Plast Reconstr Aesthet Surg 2014;67:e297–e302.
10 Natvig K, Soberg R: Relationship of intraoperative rupture of pleomorphic adenomas to recurrence: an 11–25 year follow-up study. Head Neck 1994;16: 213–217.
11 Witt RL, Eisele DW, Morton RP, et al: Etiology and management of recurrent parotid pleomorphic adenoma. Laryngoscope 2015;125:888–893.
12 Liu FF, Rotstein L, Davison AJ, et al: Benign parotid adenomas: a review of the Princess Margaret Hospital experience. Head Neck 1995;17:177–183.
13 Carew JF, Spiro RH, Singh B, et al: Treatment of recurrent pleomorphic adenomas of the parotid gland. Otolaryngol Head Neck Surg 1999;121:539–542.
14 Renehan A, Gleave EN, McGurk M: An analysis of the treatment of 114 patients with recurrent pleomorphic adenomas of the parotid gland. Am J Surg 1996; 172:710–714.
15 Samson MJ, Metson R, Wang CC, et al: Preservation of the facial nerve in the management of recurrent pleomorphic adenoma. Laryngoscope 1991;101: 1060–1062.
16 Zbaren P, Tschumi I, Nuyens M, et al: Recurrent pleomorphic adenoma of the parotid gland. Am J Surg 2005;189:203–207.
17 Speight PM, Barrett AW: Salivary gland tumours. Oral Dis 2002;8:229–240.
18 Fu KK, Leibel SA, Levine ML, et al: Carcinoma of the major and minor salivary glands: analysis of treatment results and sites and causes of failures. Cancer 1977; 40:2882–2890.
19 Shidnia H, Hornback NB, Hamaker R, et al: Carcinoma of major salivary glands. Cancer 1980;45:693–697.
20 Tu G, Hu Y, Jiang P, et al: The superiority of combined therapy (surgery and postoperative irradiation) in parotid cancer. Arch Otolaryngol 1982;108:710–713.

21 Theriault C, Fitzpatrick PJ: Malignant parotid tumors. Prognostic factors and optimum treatment. Am J Clin Oncol 1986;9:510–516.
22 Armstrong JG, Harrison LB, Spiro RH, et al: Malignant tumors of major salivary gland origin. A matched-pair analysis of the role of combined surgery and postoperative radiotherapy. Arch Otolaryngol Head Neck Surg 1990;116:290–293.
23 North CA, Lee DJ, Piantadosi S, et al: Carcinoma of the major salivary glands treated by surgery or surgery plus postoperative radiotherapy. Int J Radiat Oncol Biol Phys 1990;18:1319–1326.
24 Renehan AG, Gleave EN, Slevin NJ, et al: Clinico-pathological and treatment-related factors influencing survival in parotid cancer. Br J Cancer 1999;80:1296–1300.
25 Terhaard CH, Lubsen H, Rasch CR, et al: The role of radiotherapy in the treatment of malignant salivary gland tumors. Int J Radiat Oncol Biol Phys 2005; 61:103–111.
26 Mendenhall WM, Morris CG, Amdur RJ, et al: Radiotherapy alone or combined with surgery for salivary gland carcinoma. Cancer 2005;103:2544–2550.
27 Salgado LR, Spratt DE, Riaz N, et al: Radiation therapy in the treatment of minor salivary gland tumors. Am J Clin Oncol 2014;37:492–497.
28 Garden AS, el-Naggar AK, Morrison WH, et al: Postoperative radiotherapy for malignant tumors of the parotid gland. Int J Radiat Oncol Biol Phys 1997; 37:79–85.
29 Zeidan YH, Shultz DB, Murphy JD, et al: Long-term outcomes of surgery followed by radiation therapy for minor salivary gland carcinomas. Laryngoscope 2013;123:2675–2680.
30 Raaijmakers E, Engelen AM: Is sensorineural hearing loss a possible side effect of nasopharyngeal and parotid irradiation? A systematic review of the literature. Radiother Oncol 2002;65:1–7.
31 Leonetti JP, Marzo SJ, Zender CA, et al: Temporal bone osteoradionecrosis after surgery and radiotherapy for malignant parotid tumors. Otol Neurotol 2010;31:656–659.
32 Laramore GE, Krall JM, Griffin TW, et al: Neutron versus photon irradiation for unresectable salivary gland tumors: final report of an rtog-mrc randomized clinical trial. Radiation therapy oncology group. Medical research council. Int J Radiat Oncol Biol Phys 1993;27:235–240.
33 Schulz-Ertner D, Nikoghosyan A, Didinger B, et al: Therapy strategies for locally advanced adenoid cystic carcinomas using modern radiation therapy techniques. Cancer 2005;104:338–344.
34 Pommier P, Liebsch NJ, Deschler DG, et al: Proton beam radiation therapy for skull base adenoid cystic carcinoma. Arch Otolaryngol Head Neck Surg 2006; 132:1242–1249.
35 Cerda T, Sun XS, Vignot S, et al: A rationale for chemoradiation (vs radiotherapy) in salivary gland cancers? On behalf of the REFCOR (French Rare Head and Neck Cancer Network). Crit Rev Oncol Hematol 2014;91:142–158.

Dr. David Thomson, MA, MD, FRCR, Consultant Clinical Oncologist
The Christie NHS Foundation Trust
550 Wilmslow Rd
Manchester M20 4BX (UK)
E-Mail David.Thomson@christie.nhs.uk

Chemotherapy and Targeted Therapy

Mehmet Sen · Robin Prestwich

St. James's University Hospital, Leeds, UK

Abstract

Salivary gland cancers are uncommon neoplasms of the head and neck that exhibit considerable pathological, biological, and clinical diversity, resulting in a paucity of prospective data regarding the use of non-surgical treatments. Chemotherapy has shown limited activity in patients with metastatic disease, and there has been little exploration of its use in definitive management. There is no standard recommendation for the use of systemic therapy, with palliative chemotherapy being considered on an individual basis for rapidly progressive or symptomatic disease. Recent research has focused on the identification of characteristic molecular signatures and genomic alterations in specific histologic subtypes. Using a molecular biological approach, several signalling pathways have been identified in many cancers of salivary gland origin; thus, a number of targeted therapies have been investigated.

© 2016 S. Karger AG, Basel

Introduction

Salivary gland carcinomas (SGCs) are rare tumours. They comprise <5% of all cancers of the head and neck and encompass a wide spectrum of histological types with different biological behaviours. The standard initial therapy of localised disease consists of surgery ± adjuvant radiotherapy, and for locally unresectable disease, definitive radiotherapy is considered. The benefits of chemotherapy and other systemic treatments are uncertain, with systemic approaches usually reserved for palliative treatment of metastatic disease or loco-regional recurrence for which further surgery or radiation is not possible, as determined on an individual basis [1].

Disease Behaviours

The natural histories and biological behaviours of SGCs may be very different amongst the histological types, with the high-grade types being biologically more aggressive [anaplastic carcinoma, carcinoma ex-pleomorphic adenoma, squamous cell carcinoma, high-grade mucoepidermoid carcinoma (MEC), and salivary duct carcinoma (SDC)] than the low-grade (acinic cell carcinoma, low-grade adenocarcinoma, and polymorphous low-grade adenocarcinoma) or intermediate-grade types (adenoid cystic carcinoma). Adenocarcinoma, for example, includes a wide range of different variants and subtypes, from indolent, low-grade lesions with little tendency to spread (acinic cell carcinoma and polymorphous low-grade adenocarcinoma) to lesions with aggressive histologies [SDC and adenocarcinoma (not otherwise specified)]. Some of these tumours overexpress epidermal growth factor receptor (EGFR), HER2 and androgen receptors (ARs).

The standard treatment for local and locally advanced disease is radical surgical excision, which should be followed by adjuvant radiotherapy in the presence of high-risk features. Loco-regional recurrence is the main cause of treatment failure; select cases of loco-regional recurrence can be managed with further surgery and/or radiotherapy, although the prognosis of these patients remains poor. Their management has been reviewed in detail in previous chapters.

Patients with local regional disease who are not potentially salvageable with surgery and/or radiotherapy and those with distant metastatic disease [lungs (80%), bone (15%), and liver (5%)] will exhibit progression, highlighting the need for an effective systemic treatment including chemotherapy, targeted treatments and hormonal treatments.

Recurrent and/or Metastatic Salivary Gland Carcinomas

The natural history of metastatic disease is variable, and some patients remain asymptomatic for long time periods. This is especially true for ACC, particularly when the metastatic disease is limited to the lung [2, 3]. Although there is a wide spectrum of behaviours, the median survival time following the development of metastatic ACC is approximately 3 years, which is substantially longer than expected for most of the other SGCs [4, 5].

Treatment options for recurrent SGCs are limited by previous treatment and resectability of the loco-regionally recurrent disease. Unresectable loco-regionally recurrent disease and distant metastatic SGCs are incurable, and the goal of any active treatment is palliation. Outcomes with the use of palliative chemotherapy have been generally disappointing, with patient response rates (including partial responses) of 0–40%. Although used as a surrogate for clinical benefit, radiological responses do not necessarily translate into meaningful clinical benefit; they are generally of a short duration, with no clear evidence of improvement in progression-free survival or overall survival [6]. There are limited clinical trials that have defined the role of systemic therapy in the palliative management of SGCs. These include some phase II trials of ACC and retrospective reports of institutional experiences; however, prospective data on the other histological subtypes are scarce. These limited data suggest that there are differences in chemotherapy sensitivity among the histologic subtypes of SGCs. Currently, there are no reports of phase III trials. Therefore, watchful waiting may be the most appropriate strategy for patients with indolent disease who have few or no symptoms. Systemic therapy may be reserved for patients with symptoms and/or rapid disease progression.

The roles of systemic treatments, including chemotherapy, targeted treatments and hormonal therapy, in recurrent and metastatic SGCs have been studied and reported for the subtypes of ACC, adenocarcinoma and MEC and are discussed in the following sections.

Single-Agent Chemotherapy

Cisplatin

Cisplatin is the most commonly studied agent as monotherapy. It has been reported to provide a response rate of up to 70% in small case series of less than 15 cases [7]. In contrast with these favourable results, no objective responses were recorded in a study conducted by de Haan et al. [8] that included 10 patients with advanced ACC treated with single-agent cisplatin. In fact, according to de Haan and colleagues, the role of cisplatin in SGC remains questionable. More recently, Licitra et al. [7] performed the largest phase II study of 25 patients treated with cisplatin as monotherapy, reporting that the response rates for patients who received and those who did not receive cisplatin were 16 and 21%, respectively. For those with metastatic or loco-regional disease, the rates were 7 and 18%, respectively. Differences in responsiveness of the histologic types have been claimed, but no conclusions have been drawn due to the lack of sufficient studies. Additionally, the response duration is short, falling between 5 and 9 months.

Vinorelbine
As a single agent, vinorelbine has been reported to have some activity in ACC and adenocarcinoma [9]. It is well tolerated, and its activity is similar to those reported for cisplatin, 5-fluorouracil (5-FU), and anthracyclines.

Epirubicin
In a phase II trial including 20 patients with advanced or recurrent ACC treated with epirubicin, Vermorken et al. [10] reported a low objective response rate (10%) and rapid improvement in symptom-related disease activity in 5 patients (29.4%). However, the median time to progression was very short (16 weeks), and the median survival time was ~60 weeks.

Paclitaxel
In a phase II trial of single-agent paclitaxel conducted by the Eastern Cooperative Oncology Group on 45 patients with advanced MEC, adenocarcinoma or ACC, Gilbert et al. [11] reported a modest response rate for patients with the MEC or adenocarcinoma histologic subtype, noting 3 objective responses among 14 patients and 5 objective responses among 17 patients, respectively. No responses were observed in the ACC group. The median survival time for the entire group was 12.5 months, with a 3-year projected survival rate of 11% for the MEC patients and of 20% for the adenocarcinoma patients. A recent systematic review failed to find sufficient activity of paclitaxel in the treatment of metastatic or locally recurrent ACC [12].

Docetaxel
Based on its impressive anti-tumour activity in patients with head and neck cancer, especially squamous cell carcinoma, Raguse et al. [13] evaluated the activity of docetaxel in 4 patients with high-grade MEC of the major salivary glands. The treatment was well tolerated, with a complete response in two patients and a partial response in the other two patients. Docetaxel seems to be a logical alternative for the treatment of the MEC subtype of SGC, but at this time, only a few relevant reports exist [13, 14], and its true activity may be overestimated due to publication bias in favour of positive results.

Gemcitabine
Van Herpen et al. [15] evaluated the anti-tumour activity of gemcitabine in ACC in a phase II study including 21 patients. This therapy was well tolerated, but no objective response was reported despite disease stability for at least 6 months in 48% of the patients.

Conclusion 1
As monotherapies, most of these chemotherapeutic agents have modest activity in the treatment of advanced SGC, with limited response rates and a generally short duration of response. In addition, single-agent cisplatin does not have better efficacy than vinorelbine or epirubicin for the treatment of ACC, and it is associated with greater toxicity.

Combined-Agent Chemotherapy

Based on the limited activities of single-agent chemotherapeutic agents, multiple combinations of these agents have been developed and used for the management recurrent and metastatic salivary glands. The rationale for combination therapy is to maximise the likelihood of a response, which is particularly important in patients with significant disease-related symptoms.

Cisplatin-based combination regimens have also been explored, but the response rates have been modest, and the impact, if any, on survival has been impossible to discern. Several recent reviews have summarised these data [16, 17]. Many reported series have included large numbers of patients with ACC, reflecting the long natural history of this disease, even when metastatic. It is this long, indolent natural history, however, that makes these patients less likely to exhibit an objective disease response and makes any interpretation of the experience after chemotherapy difficult. Disease stability after a short course of chemotherapy in patients with such an indolent disease would be expected and is likely to have little clinical significance.

Cyclophosphamide/Doxorubicin/Cisplatin
The most commonly studied regimen, cyclophosphamide/doxorubicin/cisplatin (CAP), has been reported to be an active regimen in SGC. According to the combined results of many reports of small numbers of patients [18–20], the overall response rate to CAP is 46%, and a number of patients achieve complete disease remission. Based on the incidence of objective response, CAP appears to be more active than single-agent therapy; however, these results should be interpreted cautiously since they represent patients from various trials using different doses and schedules and including only a small number of patients.

Cisplatin/Doxorubicin/5-Fluorouracil
The three chemotherapeutic agents with the most evidence of single-agent activity in SGC, namely cisplatin, doxorubicin and 5-FU, have been evaluated in combination (cisplatin/doxorubicin/5-FU; PAF) by Venook et al. [21]. Seventeen patients with advanced or recurrent SGC were included in this pilot study. Only 2 patients achieved a complete response (12%) and 4 achieved a partial response (23%), for an overall response rate of 35%. The response duration was 6–15 months. This study demonstrated that PAF was tolerable as outpatient chemotherapy. Although there is no proven survival benefit of PAF, this combination is a potential option for the palliative treatment of metastatic and/or recurrent SGC, particularly adenocarcinoma.

Cyclophosphamide/Doxorubicin/Cisplatin/ 5-Fluorouracil
Multiple chemotherapy drug combinations can be employed. In a study of 17 patients conducted by Dimery et al. [22], the combination of cisplatin, doxorubicin, cyclophosphamide and 5-FU was tested. This group reached the conclusion that despite the aggressiveness of this four-drug combination, which included the perceived most active agents at their maximal dosages, as evidenced by the toxicity, the response rate was not substantially increased in comparison with those reported by other studies.

Cyclophosphamide/Cisplatin/Pirarubucin
According to the results for CAP, the combination of an anthracycline with these two alkylating agents seems to be promising. However, anthracyclines are associated with significant potential side effects, particularly cardiotoxicity. Pirarubicin is an analogue of the antineoplastic anthracycline doxorubicin that is known to be less cardiotoxic than doxorubicin and has been demonstrated to exhibit activity against some doxorubicin-resistant cell lines. Tsukuda et al. [18] reported a response rate of 36% (5/14) in a study including 14 patients with adenocarcinoma or adenocystic carcinoma of the salivary glands treated with a combination chemotherapy regimen of cyclophosphamide, pirarubicin and cisplatin. The median duration of response was 37 months in the one patient showing a complete response and 16 months (range, 6–20) in the 4 patients exhibiting partial responses. The side effects were less severe with this regimen than with CAP.

Cyclophosphamide/Doxorubicin
Posner et al. [23] reported five objective responses (38%) among 13 patients treated with the cyclophosphamide-doxorubicin regimen. Kaplan et al. [24] showed a response rate of 33% (5 partial responses among 15 patients). The overall objective response rate was 35% in both reports. This regimen was well tolerated and effective in patients with all histologic subtypes, except for MEC.

Cisplatin/Mitoxantrone
Mitoxantrone is an anthraquinone antineoplastic agent with structural and functional similarities to anthracyclines. Because it is less cardiotoxic than anthracyclines and analogous to the CAP regimen, Gedlicka et al. [25] evaluated the efficacy of the combination of cisplatin/mitoxantrone in a phase II trial involving 14 patients with recurrent or metastatic SGC. Two patients exhibited partial responses to this treatment (response durations of 27 and 14 months, respectively) yielding an overall response rate of 14.3%. With regard to tolerance, myelosuppression was commonly observed (grade 3 or 4 in 60%). In conclusion, no additional benefit has been yielded with the cisplatin/mitoxantrone combination compared to CAP, but it seems to be more myelotoxic.

Gemcitabine/Cisplatin
Laurie et al. [26] evaluated the combination of gemcitabine and cisplatin (or carboplatin) in a phase II study including 33 patients with advanced SGC. Toxicity was within the range expected for this combination, and 8 objective responses were observed (1 complete and 7 partial), for a response rate of 24%. The response durations ranged from 1.3 to 11.3 months, with a median of 6.7 months. In addition, responses were observed for all of the common histologic subtypes. This regimen may have promising activity in patients with adenocarcinoma histology. Given the absence of responses in patients treated with carboplatin in this trial due to impairment of renal function or hearing deficits, Laurie et al. [26] do not support the routine substitution of carboplatin for cisplatin for the treatment of advanced SGC. Moreover, the gemcitabine/cisplatin regimen does not offer an advantage over other cisplatin-based regimens, particularly CAP or single-agent cisplatin. In conclusion, this combination demonstrates modest activity in advanced SGC.

Conclusion 2
Combination chemotherapy regimens generally result in higher response rates compared with monotherapy; however, these schedules are generally more toxic and have not been demonstrated to provide clear benefits or to produce clinically meaningful outcomes considering their additional toxicity.

The various studies published to date are difficult to compare, and identifying the most effective chemotherapeutic regimen is not easy due to the low incidence of this disease and its rare histologies. Initial treatment with the combination of platinum and anthracycline may be preferable for patients who are very symptomatic to maximise the likelihood of a response. Otherwise, a single-agent therapy is sufficient.

Molecular Targeted Treatments

The increased understanding of the underlying molecular changes in malignant salivary gland tumours has led to the identification of several potential therapeutic targets, as shown in table 1 [2].

Few objective responses to these agents have been reported in phase II trials, but the high rates and long durations of disease stabilisation are interesting and encouraging. However, whether these findings reflect anti-tumour activity or the indolent natural history of many salivary gland tumours is unclear, particularly since many trials did not require clear evidence of progression prior to

Table 1. Frequencies of expression of molecular targets in SGC

Histology	HER2	EGFR	c-kit	Androgen receptor	Oestrogen receptor	Progesterone receptor
Adenoid cystic	Rare	Variable	Common	Rare	Rare	Rare
Adenocarcinoma	Uncommon	Uncommon	Variable	Uncommon	Rare	Rare
Mucoepidermoid	Uncommon	Common	Rare	Rare	Rare	Rare
Salivary duct	Common	Common	Rare	Common	Rare	Rare

enrolment. The current data are too preliminary to recommend the routine use of any of these agents.

Tyrosine Kinase Inhibitors

Imatinib

c-kit tyrosine kinase receptor is expressed in up to 100% of ACC cases, and it is associated with the histologically solid subtype. It is present in up to 60% of basal cell adenocarcinomas and 50% of basaloid squamous carcinomas. Seven studies have assessed the use of imatinib to treat over 80 patients with advanced ACC (many studies have used imatinib alone, and one study has evaluated the imatinib/cisplatin combination), reporting only 4 partial responses, for an objective response rate of 5% [27–29]. The durations of these responses were short (range, 9–15 months). However, stable disease was reported more commonly in 29% of patients (21 of 72 patients). In light of these poor results and the lack of c-kit exon mutations detected to date, the anti-tumour activity of imatinib in ACC remains questionable.

Gefitinib

It has been well established that the over-expression or altered expression of EGFR is involved in the pathogenesis of many tumours. It has also been reported that EGFR over-expression in ACC varies from 0 to 85% [30, 31]. A possible interaction between EGFR and SGC has been suggested. Glisson et al. [32] evaluated the efficacy of an orally active EGFR tyrosine kinase inhibitor (gefitinib) in the treatment of 28 patients with advanced SGC. The treatment was well tolerated, but no objective responses were documented. However, stable disease was observed in 14 patients (67%) with a median duration of 3 months (range, 1–4.5 months). Given the indolent nature of SGC, a stable disease response is an unreliable end-point, especially for short response durations. Thus, interpretation of the value of stable disease requires further investigation.

Lapatinib

Lapatinib, an orally active dual inhibitor of EGFR and HER2, was studied in a phase II trial of 40 patients with progressive metastatic or recurrent EGFR and/or HER2 over-expressing metastatic salivary gland tumours. The treatment was well tolerated. No objective responses were observed, but 13 patients (33%) had stable disease for at least 6 months. In this study, disease progression was required prior to enrolment; thus, the observed disease stabilisation was more likely due to lapatinib than the relatively indolent natural course of SGC [33].

Monoclonal Antibodies

Cetuximab

Another method is to target EGFR with an anti-EGFR monoclonal antibody such as cetuximab. In one study, no objective responses were observed in 30 patients (23 with ACC), although 12 of the patients with ACC had disease stabilisation for a median of 6 months, as did 3 of those with non-ACC histology [34].

Trastuzumab

Several studies have evaluated the over-expression of c-ErbB2 (HER2/neu) in SGC by either immunostaining or fluorescence *in situ* hybridisation. According to the different histologic subtypes, there is a large disparity in the proportion of SGC with HER2 over-expression. Up to one third of MECs and an even higher proportion of SDCs exhibit HER2 over-expression [35–37]. In contrast, HER2 expression is unusual in ACC and adenocarcinoma [35, 38]. In a phase II trial of trastuzumab, an anti-HER2 monoclonal antibody, that included 14 patients with advanced SGC with HER2/neu over-expression, Haddad et al. [39] identified only one partial response lasting longer than 2 years in a patient with MEC. Additionally, Nabili et al. [37] reported a complete response lasting for 3 years in one patient among three with progressive SDC treated with trastuzumab. There are also some case reports demonstrating lasting responses of SDC patients to trastuzumab [40–42]. In conclusion, the clinical utility of trastuzumab therapy in SGC of intercalated duct origin is very limited; however, this agent has shown evidence of activity in the excretory duct subtypes.

Bortezomib

In theory, the inhibition of NF-kappa-B activity can suppress the growth of SGC. Argiris et al. [43] evaluated the activity of bortezomib in a phase II trial including 25 patients with advanced ACC. The treatment was well tolerated. There were no complete or partial responses from bortezomib as monotherapy, but fifteen patients (71%) showed disease stabilisation for a median duration of 4.2 months (range, 0–20.1 months).

Sunitinib

Sunitinib, an oral inhibitor of VEGFR, PDGFR and RET, was studied in 14 patients with progressive ACC of the salivary glands. No objective responses were observed, although 5 patients had disease stabilisation for at least 6 cycles of therapy, with a median time to progression of 7.2 months. Three patients were removed from the study due to toxicity, and 10 patients required dose reductions [44].

Sorafenib

Sorafenib, a multi-targeted tyrosine kinase, was evaluated in 23 patients with advanced ACC [45]. Two objective responses were observed, with a median progression-free survival time of 11.3 months. The treatment was poorly tolerated, with over half of the patients experiencing grade 3 toxicity; the authors felt that the agent should not be further evaluated in this patient population.

Hormonal Treatment

The majority of SGCs do not express hormone receptors [46, 47]. However, it has been reported that some SGCs possess hormonal receptors, such as oestrogen [48] or progesterone receptors [49, 50] or even ARs, as in SDC [51]. In light of these reports, hormonal treatment has been used occasionally, and isolated case reports have documented objective responses in patients with ACC treated with tamoxifen [52, 53] and in those with SDC and adenocarcinoma treated with anti-androgen therapy [54, 55].

Recently, a multicentre, EORTC/UK NCRI randomised phase II study aiming to evaluate the efficacy and safety of chemotherapy versus androgen deprivation therapy in patients with recurrent and/or metastatic SGC expressing ARs has been developed, and recruitment is planned to begin by the end of 2015.

Conclusions

The treatment of inoperable and metastatic SGC remains a major challenge, with no clearly proven benefit from systemic therapy. Conventional chemotherapy, either administered as a monotherapy or combination therapy, has limited efficacy. The outcomes of targeted therapy have been broadly disappointing. Anti-androgen therapy is the focus of a forthcoming trial aiming to determine its potential role in androgen-expressing SDC. In summary, to date and in view of the modest activities of all current agents and the natural indolent behaviour of SGC, there is no evidence that systemic therapy improves survival; therefore, consideration of any systemic therapy is made on an individual basis and should be reserved for the relief of disease-related symptoms or for cases of rapid progression. The choice of any potential systemic treatment should be dictated by the histologic subtype, patient characteristics and comorbidities, toxicity and cost of the drugs.

References

1 Adelstein DJ, Koyfman SA, El-Naggar AK, et al: Biology and management of salivary gland cancers. Semin Radiat Oncol 2012;22:245–253.
2 Laurie SA, Licitra L: Systemic therepay in the palliative management of salivary gland cancers. J Clin Oncol 2006;24:2673–2678.
3 van der Wal JE, Becking AG, Snow GB, et al: Distant metastasis of adenoid cystic carcinoma of the salivary glands and the value of diagnostic examinations during the follow-up. Head Neck 2012;24:779.
4 Sur RK, Donde B, Levin V, et al: Adenoid cystic carcinoma of the salivary glands: a review of 10 years. Laryngoscope 1997;107:1276.
5 Spiro RH: Distant metastasis in adenoid cystic carcinoma of salivary gland origin. Am J Surg 1997;174:495–498.
6 Lagha A, Chraiet N, Ayadi M, et al: Systemic therapy in the management of metastatic or advanced salivary gland cancers Oral Oncology 2012;48:948–957.
7 Licitra L, Marchini S, Spinazze S, et al: Cisplatin in advanced salivary gland carcinoma: a phase II study of 25 patients. Cancer 1991;68:1874–1877.
8 de Haan LD, De Mulder PH, Vermorken JB, et al: Cisplatin-based chemotherapy in advanced adenoid cystic carcinoma of the head and neck. Head Neck 1992;14:273–277.
9 Airoldi M, Pedani F, Succo G, et al: Phase II randomised trial comparing vinorelbine versus vinorelbine plus cisplatin in patients with recurrent salivary gland malignancies. Cancer 2001;91:541–547.
10 Vermorken JB, Verweij J, de Mulder PH, et al: Epirubicin in patients with advanced or recurrent adenoid cystic carcinoma of the head and neck: a phase II study of the EORTC head and neck Cancer Cooperative Group. Ann Oncol 1993;4:785–788.
11 Gilbert J, Li Y, Pinto HA, et al: Phase II trial of taxol in salivary gland malignancies (E1394): a trial of the Eastern Cooperative Oncology Group. Head Neck 2006;28:197–204.
12 Laurie SA, Ho AL, Fury MG, et al: Systemic therapy in the management of metastatic or locally recurrent adenoid cystic carcinoma of the salivary glands: a systematic review. Lancet Oncol 2011;12:815–824.
13 Raguse J, Gath HJ, Bier J, et al: Docetaxel (taxotere) in recurrent high grade mucoepidermoid carcinoma of the major salivary glands. Oral Oncol Extra 2004;40:5–7.
14 Belli F, Di Lauro L, Zappanico A, et al: Docetaxel in the treatment of metastatic carcinoma of the salivary glands report of a case. Clin Ter 1999;150:77–79.
15 Van Herpen CM, Locati LD, Buter J, et al: Phase II study on gemcitabine in recurrent and/or metastatic adenoid cystic carcinoma of the head and neck (EORTC 24982). Eur J Cancer 2008;44:2542–2545.
16 Surakanti SG, Agulnik M: Salivary gland malignancies: the role for chemotherapy and molecular targeted agents. Semin Oncol 2008;35:309–319.
17 Vattemi E, Graiff C, Sava T, et al: Systemic therapies for recurrent and/or metastatic salivary gland cancers. Expert Rev Anticancer Ther 2008;8:393–402.
18 Tsukuda M, Kokatsu T, Ito K, et al: Chemotherapy for recurrent adeno and adenoidcystic carcinomas in the head and neck. J Cancer Res Clin Oncol 1993;119:756–758.
19 Dreyfuss AI, Clark JR, Fallon BG, et al: Cyclophosphamide, doxorubicin, and cisplatin combination chemotherapy for advanced carcinoma of salivary gland origin. Cancer 1988;47:645–648.
20 Licitra L, Cavina R, Grandi C, et al: Cisplatin, doxorubicin and cyclophosphamide in advanced salivary gland carcinoma: a phase II trial of 22 patients. Ann Oncol 1996;7:640–642.
21 Venook AP, Tseng A Jr, Meyers FJ, et al: Cisplatin, doxorubicin and 5-fluorouracil chemotherapy for salivary gland malignancies: a pilot study of the Northern California Oncology Group. J Clin Oncol 1987;5:951–955.
22 Dimery IW, Legha SS, Shirinian L, et al: Fluorouracil, doxorubicin, cyclophosphamide and cisplatin combination chemotherapy in advanced or recurrent salivary gland carcinoma. J Clin Oncol 1990;8:1056–1062.
23 Posner MR, Ervin TJ, Weichselbaum RR, et al: Chemotherapy for advanced salivary gland neoplasms. Cancer 1982;50:2261–2264.
24 Kaplan MJ, Johns ME, Cantrell RW: Chemotherapy for salivary gland cancer. Otolaryngol Head Neck Surg 1986;95:165–170.
25 Gedlicka C, Schull B, Formanek M, et al: Mitoxantrone and cisplatin in recurrent and/or metastatic salivary gland malignancies. Anticancer Drugs 2002;13:491–495.

26 Laurie SA, Siu LL, Winquist E, et al: A phase II study of platinium and gemcitabine in advanced salivary gland cancers: a trial of the NCIC clinical trials group. Cancer 2010;116:362–368.
27 Faivre S, Raymond E, Casiraghi O, et al: Imatinib mesylate can induce objective response in progressing, highly expressing KIT adenoid cystic carcinoma of the salivary glands. J Clin Oncol 2005;23:6271–6273.
28 Gibbons MD, Manne U, Carroll WR, et al: Molecular differences in mucoepidermoid carcinoma and adenoid cystic carcinoma of the major salivary glands. Laryngoscope 2001;111:1373–1378.
29 Katopodi E, Patsouris E, Papanikolaou V, et al: Immunohistochemical detection of epidermal growth factor and its receptor in salivary gland carcinomas. Oral Surg Oral Med Oral Pathol Oral Radiol Endod 2003;95:266–268.
30 Vered M, Braunstein E, Buchner A: Immunohistochemical study of epidermal growth factor receptor in adenoid cystic carcinoma of salivary gland origin. Head Neck 2002;24:632–636.
31 Dodd RL, Slevin NJ: Salivary gland adenoid cystic carcinoma: a review of chemotherapy and molecular therapies. Oral Oncol 2006;42:759–769.
32 Glisson B, Blumenschein G, Francisco M, et al: Phase II trial of gefitinib in patients with incurable salivary gland cancer. J Clin Oncol 2005;23:580S (abstract 5532).
33 Agulnik M, Cohen EW, Cohen RB, et al: Phase II study of lapatinib in recurrent or metastatic epidermal growth factor receptor and/or erbB2 expressing adenoid cystic carcinoma and nonadenoid cystic carcinoma malignant tumors of the salivary glands. J Clin Oncol 2007;25:3978–3984.
34 Locati LD, Bossi P, Perrone F, et al: Cetuximab in recurrent and/or metastatic salivary gland carcinomas: a phase II study. Oral Oncol 2009;45:574–578.
35 Glisson B, Colevas AD, Haddad R, et al: HER2 expression in salivary gland carcinomas: dependence on histological subtype. Clin Cancer Res 2004;10:944–946.
36 Jaehne M, Roeser K, Jaekel T, et al: Clinical and immunohistologic typing of salivary duct carcinoma: a report of cases. Cancer 2005;103:2526–2533.
37 Nabili V, Tan JW, Bhuta S, et al: Salivary duct carcinoma: a clinical and histologic review with implications for trastuzumab therapy. Head Neck 2007;29:907–912.
38 Dori S, Vered M, David R, et al: HER2/neu expression in adenoid cystic carcinoma of salivary gland origin: an immunohistochemical study. J Oral Pathol Med 2002;31:463–467.
39 Haddad R, Colevas AD, Krane JF, et al: Herceptin in patients with advanced or metastatic salivary gland carcinomas. A phase II study. Oral Oncol 2003;39:724–727.
40 Prat A, Parera M, Reyes V, et al: Successful treatment of pulmonary metastatic salivary ductal carcinoma with trastuzumab-based therapy. Head Neck 2008;30:680–683.
41 Nashed M, Casasola RJ: Biological therapy of salivary duct carcinoma. J Laryngol Otol 2009;123:250–252.
42 Kaidar-Person O, Billan S, Kuten A: Targeted therapy with trastuzumab for advanced salivary ductal carcinoma: case report and literature review. Med Oncol 2012;29:704–706.
43 Argiris A, Ghebremichael M, Burtness B, et al: A phase 2 trial of bortezomib followed by the addition of doxorubicin at progression in patients with recurrent or metastatic adenoid cystic carcinoma of the head and neck: a trial of the Eastern Cooperative Oncology Group (E1303). Cancer 2011;117:3374–3382.
44 Chau NG, Hotte SJ, Chen EX, et al: A phase II study of sunitinib in recurrent and /or metastatic adenoid cystic carcinoma (ACC) of the salivary glands; current progress and challenges in evaluating molecularly targeted agents in ACC. Ann Oncol 2012;23:1562.
45 Thomson DJ, Silva P, Denton K, et al: Phase II trial of sorafenib in advanced salivary adenoid cystic carcinoma of the head and neck. Head Neck 2015;37:182.
46 Miller AS, Hartman GG, Chen SY, et al: Estrogen receptor assay in polymorphous low-grade adenocarcinoma and adenoid cystic carcinoma of salivary gland origin. An immunohistochemical study. Oral Surg Oral Med Oral Pathol 1994;77:36–40.
47 Pires FR, Perez DE, Almeida OP, et al: Estrogen receptor expression in salivary gland mucoepidermoid carcinoma and adenoid cystic carcinoma. Pathol Oncol Res 2004;10:166–168.
48 Dimery IW, Jones LA, Verjan RP, et al: Estrogen receptors in normal salivary gland and salivary gland carcinoma. Arch Otolaryngol Head Neck Surg 1987;113:1082–1085.
49 Jeannon JP, Soames JV, Bell H, et al: Immunohistochemical detection of oestrogen and progesterone receptors in human salivary gland tumours. Clin Otolaryngol 1999;24:52–54.
50 Ozono S, Onozuka M, Sato K, et al: Immunohistochemical localization of estradiol, progesterone, and progesterone receptor in human salivary glands and salivary adenoid cystic carcinomas. Cell Struct Funct 1992;17:169–175.
51 Fan CY, Melhem MF, Hosal AS, et al: Expression of androgen receptor, epidermal growth factor receptor, and transforming growth factor alpha in salivary duct carcinoma. Arch Otolaryngol Head Neck Surg 2001;127:1075–1079.
52 Shadaba A, Gaze MN, Grant HR: The response of adenoid cystic carcinoma to tamoxifen. J Laryngol Otol 1997;111:1186–1189.
53 Elkin AD, Jacobs CD: Tamoxifen for salivary gland adenoid cystic carcinoma: a report of two cases. J Cancer Res Clin Oncol 2008;134:1151–1153.
54 Locati LD, Quattrone P, Bossi P, et al: A complete remission with androgen deprivation therapy in a recurrent androgen receptor-expressing adenocarcinoma of the parotid gland. Ann Oncol 2003;14:1327–1328.
55 Van der Hulst RW, Van Krieken JH, Van der Kwast TH, et al: Partial remission of parotid gland carcinoma after goserelin. Lancet 1994;344:817.

Mehmet Sen, FRCR, Consultant Clinical Oncologist & Honorary Senior Lecturer
St. James's Institute of Oncology
Beckett Street
Leeds LS9 7TF (UK)
E-Mail mehmet.sen@nhs.net

Management of Inoperable Malignant Neoplasms

Ana P. Kiess · Harry Quon

Departments of Radiation Oncology and Molecular Radiation Sciences, and Otorhinolaryngology – Head and Neck Surgery, Johns Hopkins University, Baltimore, Md., USA

Abstract

For patients with inoperable salivary gland malignancy, radiation therapy has significant limitations but has been the mainstay of treatment. With standard photon radiation (X-rays), the 10-year loco-regional control (LRC) and overall survival rates are only ∼25%. Neutron radiation has potential biological advantages over photon radiation because it causes increased DNA damage, and studies of patients with inoperable salivary gland malignancy have shown improved 6-year LRC and overall survival of ∼60%. However, neutron radiation may also increase the risk of late toxicities, especially central nervous system toxicities after treatment of tumors involving the base of the skull. Proton radiation has potential physical advantages due to minimal exit dose through normal tissues, and a recent study has demonstrated 90% 5-year LRC after combined proton/photon radiation for adenoid cystic carcinoma involving the base of the skull. Stereotactic radiosurgery has also been used in combination with neutrons or standard photons as a technique to boost the skull base. The use of concurrent chemotherapy as a radiosensitizer has been considered based on extrapolation of data on squamous cell carcinomas, but further data are needed on inoperable salivary gland malignancies. Newer targeted therapies are also under investigation, and clinical trial enrollment is encouraged.

© 2016 S. Karger AG, Basel

Introduction

Only about 8% of patients with salivary gland tumors present with inoperable disease, and there is a lack of robust clinical data on this rare and challenging condition [1]. The available data suggest the limited efficacy of standard photon (X-ray) radiotherapy (RT) and the possible improved efficacies of neutron, proton, and carbon ion RT, which are only available at few medical centers. Therefore, even experts in salivary gland malignancies find it difficult to provide patients with clear treatment recommendations.

This chapter addresses the management of patients with newly diagnosed inoperable salivary gland malignancy. The disease patterns are discussed, as well as comprehensive management from initial evaluation to specific treatment techniques and follow-up. Summary treatment recommendations are provided, and areas of importance for further study are also highlighted. Other chapters address the indications for adjuvant radiation therapy [discussed further in the chapter by Thomson et al., this vol., pp. 141–147], the use of chemotherapy and targeted therapy [discussed further in the chapter by Sen and Prestwich, this vol., pp. 148–156], and the management of recurrent loco-regional disease [discussed further in the chapter by Merdad et al., this vol., pp. 168–174]. There is some overlap with regard to recurrent loco-regional disease, as many studies of inoperable salivary gland tumors have included patients with either upfront or recurrent inoperable disease. In addition, there is some discussion on the use of chemotherapy for inoperable disease.

Patterns of Disease

Inoperable salivary gland tumors are typically of clinical stage T4b according to the TNM Classification of Malignant Tumors of the Union for International Cancer Control 7th Edition [discussed further in the chapter by Bradley, this vol. pp. 1–8]. Stage T4b includes tumors with invasion of the skull base or pterygoid plates or encasement of the carotid artery. However, with current surgical techniques, some tumors with these characteristics are now considered resectable by expert teams. The current criteria for inoperability are beyond the scope of this chapter but may include the following: facial nerve spread beyond the mastoid segment; trigeminal nerve spread to the cavernous sinus, foramen ovale or Meckel's cave; carotid artery encasement, especially at the skull base; bony invasion of the clivus; involvement of the prevertebral fascia; and/or brain invasion [2, 3]. The radiologic criteria for carotid artery encasement include greater than 270 degrees of circumferential tumor contact, deformation of the artery, and segmental obliteration of the fat around the artery [4, 5].

Perineural spread (PNS) along the cranial nerves to the skull base is the most common reason for inoperability in salivary gland malignancies, and this is especially common in adenoid cystic carcinoma (ACC), in which perineural invasion is seen histologically in 50% of cases [6].

PNS can occur in both directions along the nerve but typically travels in a contiguous retrograde fashion toward the brainstem. 'Skip' lesions have been reported but may be related to the branching patterns of involved nerves rather than true discontinuity of spread [7]. For primary parotid malignancies, PNS most commonly occurs along the facial nerve (VII) through the stylomastoid foramen and temporal bone (fig. 1b). The auriculotemporal nerve can also be a conduit for PNS from the parotid gland to the mandibular nerve (V3) and foramen ovale (fig. 1a). Minor salivary gland malignancies of the hard palate often spread along the palatine nerves through the greater and lesser palatine foramina to join V2 in the pterygopalatine fossa. Primary submandibular gland malignancies can spread along the lingual nerve (branch of V3) and/or the hypoglossal nerve (XII).

Clinical Manifestations and Patient Evaluation

Most patients with early-stage salivary gland malignancy present with a solitary painless mass, but patients with advanced-stage, inoperable disease often have additional symptoms related to local spread and PNS. Proper evaluation begins with a detailed history and physical examination, including head and neck examination, fiberoptic nasopharyngolaryngoscopy and cranial nerve examination. For primary parotid malignancies, facial twitching and/or paresis due to PNS are often initially misdiagnosed as viral Bell's palsy, as was the case for the patient presented in figure 1. Symptoms of trigeminal nerve PNS may include facial paresthesia, numbness, or pain, which are often initially misdiagnosed as benign trigeminal neuralgia. Other symptoms of PNS may include tongue paresis (hypoglossal nerve), dysphagia (glossopharyngeal nerve) and vocal fold paresis (vagus nerve). Symptoms of local invasion may include pain, trismus (due to invasion of the parapharyngeal space or pterygoids), or dysphagia due to a mass effect.

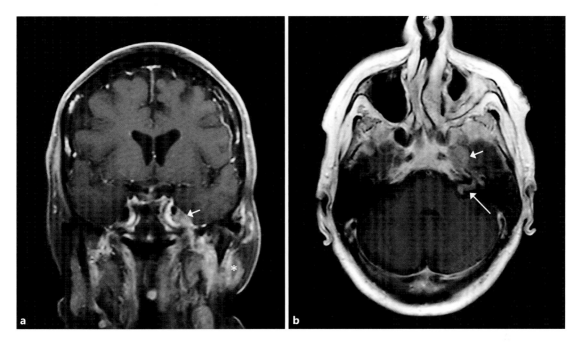

Fig. 1. T1-weighted post-contrast MRI from a 71-year-old woman who presented with multiple left-sided cranial nerve palsies (VII, V3, IX, and X). The images show a deep enhancing parotid mass (asterisk in **a**) with perineural spread along the auriculotemporal nerve (branch of V3) through the foramen ovale to Meckel's cave (short white arrows in **a** and **b**), as well as along the facial nerve (VII) to the internal auditory canal (long white arrow in **b**). Fine-needle aspiration of the left parotid mass showed high-grade adenocarcinoma.

Imaging evaluation of advanced salivary gland malignancy usually includes both contrast-enhanced CT and MRI from the base of skull to the clavicles [8–10]. MRI is essential for delineation of soft tissue invasion, PNS, and intracranial extension (fig. 1). CT is helpful for evaluation of bony destruction and regional lymph node metastases. Both of these modalities are used by the surgical team to determine whether the disease is resectable [2]. Chest imaging is also recommended for patients with advanced disease [10]. Positron emission tomography is not well established for salivary gland malignancies, but early studies have demonstrated its use to detect unrecognized nodal and distant metastases [11, 12].

Other important elements of the initial patient evaluation include plastic surgery evaluation for facial paresis, speech pathology and nutritional evaluations for dysphagia, laryngology evaluation for vocal cord paresis, and dental evaluation prior to RT.

Conventional Photon Radiation

The current guidelines of the National Comprehensive Cancer Network (NCCN 1.2015) for inoperable salivary gland malignancies advise treatment with conventional photon or photon/electron RT using conformal techniques, with delivery of 66–70 Gy to gross disease and high-risk areas in 1.8–2.0 Gy fractions [10]. Unfortunately, this treatment technique has shown limited locoregional control (LRC) rates of 17–30% for patients with stage T4 cancer in most retrospective studies and one prospective randomized trial [1, 13–15]. Pooled analysis of inoperable and recurrent disease has shown a local control (LC) rate of 25% with photons and/or electrons [16].

In a retrospective series from the University of Florida of 204 patients with salivary gland carcinoma, a high T stage and lack of surgery were predictive of local recurrence, and the use of RT alone for patients with stage T4 cancer resulted in only 21% LC and 21% overall survival (OS) at 10 years [15]. Severe late toxicities were seen in 6% of the patients, including osteoradionecrosis, anticipated unilateral vision loss, and fistula. A study from the University of California at San Francisco of 45 patients treated with RT alone showed that those with stage T4 cancer had a 10-year LC rate of 30% and an OS of 37% [13]. As expected, a high T stage was predictive of local recurrence, as well as an inadequate RT dose of less than 66 Gy. In a large Dutch study of 538 patients with salivary gland malignancy, LC after primary radiation showed a clear dose-response relationship, with 50% LC at 5 years (for all stages) if doses of >66 Gy were used compared with 0% if doses of <66 Gy were used (p = 0.0007) [1]. An international randomized trial of photon versus neutron radiation for patients with inoperable salivary gland malignancy again showed poor results for photon radiation, with only 17% LC at 10 years [14]. This trial is discussed in detail below. Standard photon radiation techniques are also described below, as well as the potential roles of concurrent chemotherapy, stereotactic radiation and hyperfractionated radiation.

Neutron and Carbon Ion Radiation

Neutrons and other heavy particles such as carbon ions have a biological advantage over photons because they induce more frequent double-stranded DNA breaks along their path. This type of damage is less dependent on oxygenation and the cell cycle than the more sparse DNA damage caused by photons. In radiobiologic terms, neutrons have higher linear energy transfer (LET; i.e. more energy transferred to the tissue per path length), a lower oxygen enhancement ratio (i.e. less radioresistance with hypoxia), and higher relative biologic effectiveness (i.e. increased tumor response for a given dose). Salivary gland malignancies typically have a low growth fraction and a long doubling time, making high-LET radiation particularly favorable.

A randomized trial was conducted by the Radiation Therapy Oncology Group (RTOG) and Medical Research Council (MRC) comparing neutron radiation to photon/electron radiation in patients with inoperable primary and recurrent malignant salivary gland tumors (fig. 2) [14]. The patients received either daily photon/electron radiation for 4–7 weeks (55–70 Gy depending on the institutional fractionation) or 12 fractions of neutron radiation over 4 weeks (16–22 nGy depending on the institutional relative biologic effectiveness). The trial was stopped early due to improved outcomes achieved with neutrons, and final analysis of 25 patients at 10 years showed significantly improved LC with neutrons (56%) versus photons (17%; p < 0.005). There was initially a trend toward an improved 2-year OS (62 vs. 25%, p = 0.1), but with longer follow-up time, there was no significant difference in OS. Distant metastases accounted for the majority of failures in the neutron arm, whereas loco-regional failures accounted for the majority in the photon arm. Other institutional reports using neutron radiation have also shown similarly good LC rates of 55–75%, particularly in patients with ACC [17–24]. A large recent update from the University of Washington reported a 6-year LRC rate of 59% for 279 patients with salivary gland tumors treated with neutrons (94% with gross disease) [24].

There have been concerns with neutron RT regarding severe late toxicities, especially central nervous system (CNS) toxicities in patients with

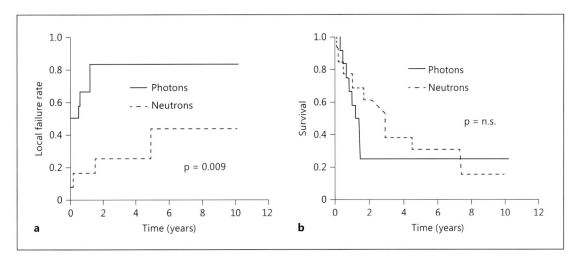

Fig. 2. a A randomized trial comparing photon and neutron RT was stopped early (after 25 patients) due to significantly higher local failure in the photon arm. **b** There was no difference in overall survival with long-term follow-up. Reproduced with permission from Laramore et al. [14].

skull base involvement [17, 24–26]. In the RTOG/MRC trial, 9 out of 13 patients in the neutron arm developed severe late toxicity compared to only 4 out of 12 in the photon arm [14]. In a study from the University of Washington, 10% of patients developed grade 3 or 4 toxicity, including 4 patients with CNS radionecrosis, 3 with optic neuritis causing blindness, 2 with retinopathy causing blindness, and 1 with cervical myelopathy [24]. These effects are related to the sensitivity of CNS structures to high-LET radiation, although interestingly, there did not appear to be a significant rate of facial nerve palsy as a complication of neutron RT [18, 19, 24].

Stereotactic radiosurgery (SRS) has been used in combination with neutrons as a method to safely boost the base of the skull [27]. At the University of Washington, 34 patients with skull base involvement received neutron therapy limited to 12 nGy at the skull base, followed by a Gamma Knife SRS boost of 12 Gy in one fraction, resulting in a 40-month LC rate of 82% compared with the historical LC rate of 39% with neutrons alone.

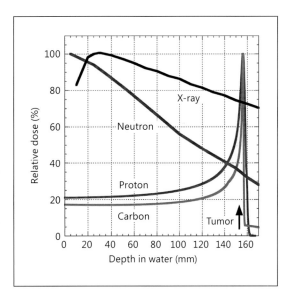

Fig. 3. Typical depth-dose distributions for different radiation techniques. Photons (X-rays) and neutrons gradually decrease in dose with depth, whereas carbon ions and protons have sharp peaks in dose at specific depths, with minimal exit doses beyond the tumor. Reproduced with permission from Fukumara et al. [28].

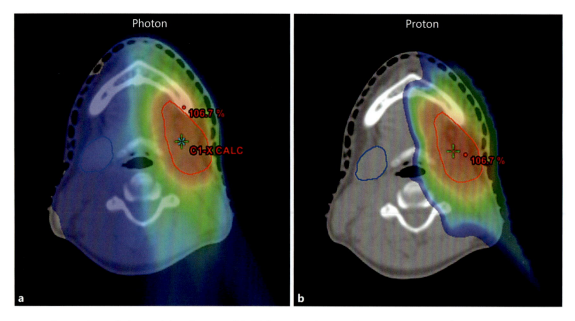

Fig. 4. Comparison of photon (**a**) and proton (**b**) RT dose distributions for a patient with left submandibular gland adenoid cystic carcinoma. Proton RT results in minimal exit dose beyond the tumor, sparing the normal tissues. The tumor volume is red, the right submandibular gland is blue, and the colorwash minimum is 7 Gy. In this example, intensity-modulated radiation was used for photons, whereas double scattering was used for protons. Reproduced with permission from Lin et al. [31].

There were only four local failures with SRS boost, including two in field and two out of field. There were no reported increases in toxicity, with only one case of symptomatic brain necrosis [27].

As noted above, carbon ions also have a potential biological advantage due to high LET, as well as a potential physical advantage due to the sharp peak of the depth-dose curve with minimal exit dose through normal tissues (fig. 3) [28]. A recent German study of the combination of carbon ion boost and standard photon radiation for ACC showed a 4-year LRC rate of 78%, with severe late toxicities observed in less than 5% of patients [29].

In summary, current data suggest LC benefits of neutron and carbon ion RT for inoperable salivary gland malignancies, but additional data are needed. The lack of survival benefit observed in the RTOG/MRC randomized trial, the concern for severe late toxicities, and the limited availability of these therapies have significantly restricted their use to date. Neutron therapy is currently only available at three sites in the United States (Seattle, Wash., Detroit, Mich., and Batavia, Ill.), and carbon ion therapy is not available in the Unites States. However, for patients with inoperable salivary gland malignancy who are able to relocate for treatment, neutron therapy (with SRS boost for skull base disease) should be considered, preferably in a clinical trial.

Proton Radiation

The use of proton radiation has potential physical advantages over photons and neutrons, as the depth-dose curve has a sharp peak with minimal exit dose (similar to carbon ions; fig. 3) [28, 30, 31]. This may allow for the sparing of normal tissues (fig. 4), as well as possible dose escalation to disease at the skull base [28, 31]. A comparative treatment planning study in the head and neck has demonstrated that intensity-modulated *proton* RT has more potential for sparing normal tissues than intensity-modulated *photon* RT [32]. A recent study of combined proton and photon RT in 23 patients with ACC with skull base extension showed a 5-year LC rate of 92% and an OS of 77% [33]. The patients had undergone biopsy alone (48%), partial resection (39%), or gross total resection with positive margins (13%), and the median dose was 76 cobalt-gray equivalent. Another study of 26 patients with ACC (77% with skull base extension) treated with protons showed a 2-year LC rate of 95%, but longer follow-up is needed, and most of these patients had undergone prior resection [34]. Additional prospective research is necessary to investigate the potential benefit of protons over photons for inoperable salivary gland malignancies, especially as proton radiation is becoming more widely available in the United States. For patients with ACC with skull base involvement, proton therapy should be considered, preferably as part of a clinical trial.

Hypofractionated Stereotactic Photon Radiation

Hypofractionated stereotactic body radiotherapy or radiosurgery (SRS) has been proposed as a technique to safely boost skull base disease after neutron therapy (as described above) or conventional photon therapy [27, 35]. It has also been used as a single modality for re-irradiation of recurrent salivary gland malignancies at the skull base [36]. In a study of 184 patients with head and neck cancer treated with SRS (51% after conventional photon RT), 8% had severe late toxicities, including temporal lobe necrosis, cranial nerve palsy, facial numbness, pain and decreased vision [35]. Out of the 50 patients in this study with salivary gland malignancy, 11 (22%) had marginal failure. Another study of the use of SRS in 8 patients with recurrent salivary gland malignancy at the skull base showed palliation of pain in all patients and a median LC time of 15 months, with 1 patient developing radionecrosis [36].

Hyperfractionated Photon Radiation

A study conducted by Wang and Goodman [37] investigated the use of accelerated hyperfractionated photon RT in 24 patients with inoperable salivary gland malignancy. The fractionation was 1.6 Gy twice daily to a total of 65–70 Gy. For parotid cancers, they reported 100% LC and 65% OS at 5 years. For minor salivary gland cancers, they reported 78% LC and 93% OS at 5 years. However, long-term follow-up was not reported.

Concurrent Chemoradiation

The use of concurrent chemotherapy as a radiosensitizer for inoperable salivary gland malignancies is based on a low level of evidence, and the rationale for using platinum agents has been largely extrapolated from data on head and neck squamous cell carcinoma (HNSCC) and recurrent or metastatic salivary gland malignancies [38]. For advanced HNSCC, a number of randomized trials and a meta-analysis have shown a survival benefit of adding concurrent cisplatin to definitive radiation, but most of these trials have excluded patients with salivary gland malignancy [39]. Several recent retrospective studies have investigated the use of concurrent chemoradiation (with standard photon RT) specifically for salivary gland malignancies [40–44]. In a single-institution study from Japan, 17 patients with stages III–IV salivary gland carcinoma were treated with concurrent chemoradiation with cyclophosphamide, pirarubicin, and cisplatin [44]. The response rate was 76%, the complete response rate was 24%, and the 5-year OS was 70%. Common toxicities included neutropenia, leukopenia and mucositis [44]. In a multi-institutional study from the United States, 16 patients with unresected ACC were treated with concurrent chemoradiation with cisplatin or carboplatin (10 intravenous and 6 intraarterial) [42]. After a median follow-up time of 61 months, 7 patients had no evidence of disease, 5 had distant recurrence, and 3 had local progression. The 5-year local progression-free survival was 61%, and the OS was 87%. Another small study of 6 patients with inoperable parotid gland malignancy used concurrent chemoradiation with cisplatin followed by adjuvant chemotherapy with cisplatin and VP-16, showing long-term complete responses in 3 of 6 patients [43]. Given the low level of available data, the current NCCN guidelines consider concurrent systemic therapy with radiation as a category 2B recommendation for inoperable salivary gland malignancies [10].

In the future, targeted therapies and immunotherapies may provide more effective systemic responses and may have potential for synergistic interactions with RT. Salivary gland malignancies have been shown to express various molecular targets, including epidermal growth factor receptor (EGFR), human epidermal growth factor receptor 2 (HER2), c-kit, androgen receptors, and vascular endothelial growth factor [45]. A number of recent studies have investigated drugs targeting these molecules in patients with recurrent or metastatic salivary gland malignancy [discussed further in the chapter by Sen and Prestwich, this vol., pp. 148–156], and enrollment in clinical trials is encouraged for patients with inoperable disease.

Palliation

For patients with inoperable salivary gland malignancy with a poor performance status or significant co-morbidities, palliative chemotherapy, palliative RT, and hospice should also be considered [discussed further in the chapter by Glazer and Shuman, this vol., pp. 182–188]. Phase II studies of the use of palliative chemotherapy for recurrent or metastatic disease have shown response rates of 15–50% for a typical duration of 6–9 months [discussed further in the chapter by Sen and Prestwich, this vol., pp. 148–156] [45]. Palliative radiation may also be utilized in patients with inoperable disease with a variety of short fractionation options, as well as hypofractionated stereotactic body radiotherapy or SRS, as described above.

Techniques of Radiation

Intensity-modulated RT is our preferred technique in treatment planning for inoperable salivary gland malignancies, given the frequent proximity of critical normal structures, such as the brainstem, cochlea, optic chiasm, and temporal lobe of the brain. The recommended doses are 70 Gy for gross disease and 66 Gy for high-risk areas, delivered in 1.8–2.0 Gy fractions. The typical elective nodal irradiation (ENI) dose is 50–60 Gy, delivered in 1.6–2.0 Gy fractions, although ENI has not been well studied, as noted below. The contralateral major salivary glands should be limited to <10% of the prescription dose to reduce xerostomia. It is essential to include the pathway of the relevant cranial nerves from the tumor to the skull base in contouring radiation target volumes for inoperable salivary gland malignancies with PNS. Several recent contouring guides have been published addressing the cranial nerves VII, V2, and V3 [46, 47].

Nodal irradiation is indicated for all patients with clinically involved nodes. However, elective nodal irradiation is not standardized for clinically node-negative salivary gland malignancies. In general, the risk of lymph node metastasis is higher in patients with T3–T4 disease, a pharyngeal or tongue primary tumor site, and a histology of high-grade mucoepidermoid carcinoma or adenocarcinoma [48]. Based on available data [49, 50], we recommend that elective nodal irradiation should always be considered for patients with inoperable disease.

Treatment Follow-Up

Patients with inoperable salivary gland malignancy are at high risk of developing loco-regional recurrence as well as distant metastases. However, the disease may be protracted, with late recurrences occurring many years after treatment. Post-treatment baseline imaging of the primary tumor and neck (usually MRI) is recommended at approximately 3 months after treatment. However, further re-imaging and chest imaging are not routinely recommended unless indicated based on symptoms. Speech, hearing, swallow, and dental evaluations are often clinically indicated during follow-up. Follow-up visits are recommended every 1–3 months for the first year, every 2–6 months for the second year, every 4–8 months for years 3–5, and then annually thereafter [10].

Conclusions

Inoperable salivary gland malignancies present unique and challenging clinical situations. Patients often present with advanced PNS involving the skull base with associated cranial neuropathies. Conventional photon radiation is the standard of care but has poor LC and OS rates of ~25% at 10 years. Improved LC may be possible with neutron, proton, or carbon ion radiation performed at specialized medical centers, and neutron radiation has shown improved LC and OS rates of ~60% at 10 years in a small randomized trial. However, many patients are unable to relocate for treatment, and further data are needed to assess the relative benefits and risks of these treatments, particularly the risk of late CNS toxicities. We encourage consideration of neutron RT (with SRS boost for skull base involvement) or proton therapy (particularly for ACC with skull base involvement), preferably as part of a clinical trial. Further study of concurrent systemic therapies with RT should be pursued, as the current recommendations have been mainly extrapolated from data on HNSCC. Patients with inoperable salivary gland malignancy may have more treatment options in the near future as proton RT is becoming increasingly available in the United States and elsewhere and potential molecular targets are being identified for systemic therapy.

References

1 Terhaard CH, Lubsen H, Rasch CR, et al: The role of radiotherapy in the treatment of malignant salivary gland tumors. Int J Radiat Oncol Biol Phys 2005; 61:103–111.
2 Yousem DM, Gad K, Tufano RP: Resectability issues with head and neck cancer. AJNR Am J Neuroradiol 2006;27:2024–2036.
3 Hanna EY, DeMonte F (eds): Comprehensive Management of Skull Base Tumors. Boca Raton, CRC Press, 2008.
4 Pons Y, Ukkola-Pons E, Clément P, et al: Relevance of 5 different imaging signs in the evaluation of carotid artery invasion by cervical lymphadenopathy in head and neck squamous cell carcinoma. Oral Surg Oral Med Oral Pathol Oral Radiol Endod 2010;109:775–778.
5 Yousem DM, Hatabu H, Hurst RW, et al: Carotid artery invasion by head and neck masses: prediction with MR imaging. Radiology 1995;195:715–720.
6 Barrett A, Speight PM: Perineural invasion in adenoid cystic carcinoma of the salivary glands: a valid prognostic indicator? Oral Oncol 2009;45:936–940.

7 Kayahara M, Nakagawara H, Kitagawa H, et al: The nature of neural invasion by pancreatic cancer. Pancreas 2007;35: 218–223.
8 Burke CJ, Thomas RH, Howlett D: Imaging the major salivary glands. Br J Oral Maxillofac Surg 2011;49:261–269.
9 Thoeny HC: Imaging of salivary gland tumours. Cancer Imaging 2007;7:52–62.
10 Pfister DG, Spencer S, Brizel DM, et al: Head and Neck Cancers, Version 1.2015. J Natl Compr Canc Netw 2015;13:847–856.
11 Jeong HS, Chung MK, Son YI, et al: Role of 18F-FDG PET/CT in management of high-grade salivary gland malignancies. J Nucl Med 2007;48:1237–1244.
12 Razfar A, Heron DE, Branstetter BF, et al: Positron emission tomography-computed tomography adds to the management of salivary gland malignancies. Laryngoscope 2010;120:734–738.
13 Chen AM, Bucci MK, Quivey JM, et al: Long-term outcome of patients treated by radiation therapy alone for salivary gland carcinomas. Int J Radiat Oncol Biol Phys 2006;66:1044–1050.
14 Laramore GE, Krall JM, Griffin TW, et al: Neutron versus photon irradiation for unresectable salivary gland tumors: final report of an RTOG-MRC randomized clinical trial. Radiation Therapy Oncology Group. Medical Research Council. Int J Radiat Oncol Biol Phys 1993;27:235–240.
15 Mendenhall WM, Morris CG, Amdur RJ, et al: Radiotherapy alone or combined with surgery for salivary gland carcinoma. Cancer 2005;103:2544–1550.
16 Koh WJ, Laramore G, Griffin T, et al: Fast neutron radiation for inoperable and recurrent salivary gland cancers. Am J Clin Oncol 1989;12:316–319.
17 Battermann JJ, Mijnheer BJ: The Amsterdam fast neutron therapy project: a final report. Int J Radiat Oncol Biol Phys 1986;12:2093–2099.
18 Buchholz TA, Laramore GE, Griffin BR, et al: The role of fast neutron radiation therapy in the management of advanced salivary gland malignant neoplasms. Cancer 1992;69:2779–2788.
19 Catterall M, Errington RD: The implications of improved treatment of malignant salivary gland tumors by fast neutron radiotherapy. Int J Radiat Oncol Biol Phys 1987;13:1313–1318.
20 Douglas JG, Laramore GE, Austin-Seymour M, et al: Treatment of locally advanced adenoid cystic carcinoma of the head and neck with neutron radiotherapy. Int J Radiat Oncol Biol Phys 2000;46: 551–557.
21 Douglas JG, Lee S, Laramore GE, et al: Neutron radiotherapy for the treatment of locally advanced major salivary gland tumors. Head Neck 1999;21:255–263.
22 Huber PE, Debus J, Latz D, et al: Radiotherapy for advanced adenoid cystic carcinoma: neutrons, photons or mixed beam? Radiother Oncol 2001;59:161–167.
23 Saroja KR, Mansell J, Hendrickson FR, et al: An update on malignant salivary gland tumors treated with neutrons at Fermilab. Int J Radiat Oncol Biol Phys 1987;13:1319–1325.
24 Douglas JG, Koh WJ, Austin-Seymour M, et al: Treatment of salivary gland neoplasms with fast neutron radiotherapy. Arch Otolaryngol Head Neck Surg 2003;129:944–948.
25 Hong CS, Gokozan HN, Otero JJ, et al: Delayed cerebral radiation necrosis after neutron beam radiation of a parotid adenocarcinoma: a case report and review of the literature. Case Rep Neurol Med 2014;2014:717984.
26 Manz HJ, Woolley PV 3rd, Ornitz RD: Delayed radiation necrosis of brainstem related to fast neutron beam irradiation: case report and literature review. Cancer 1979;44:473–479.
27 Douglas JG, Goodkin R, Laramore GE: Gamma knife stereotactic radiosurgery for salivary gland neoplasms with base of skull invasion following neutron radiotherapy. Head Neck 2008;30:492–496.
28 Fukumura A, Tsujii H, Kamada T, et al: Carbon-ion radiotherapy: clinical aspects and related dosimetry. Radiat Protect Dosim 2009;137:149–155.
29 Schulz-Ertner D, Nikoghosyan A, Didinger B, et al: Therapy strategies for locally advanced adenoid cystic carcinomas using modern radiation therapy techniques. Cancer 2005;104:338–344.
30 Ahn PH, Lukens JN, Teo BK, et al: The use of proton therapy in the treatment of head and neck cancers. Cancer J 2014; 20:421–426.
31 Lin A, Swisher-McClure S, Bonner Millar L, et al: Proton therapy for head and neck cancer: current applications and future directions. Transl Cancer Res 2012;1:255–263.
32 Steneker M, Lomax A, Schneider U: Intensity modulated photon and proton therapy for the treatment of head and neck tumors. Radiother Oncol 2006;80: 263–267.
33 Pommier P, Liebsch NJ, Deschler DG, et al: Proton beam radiation therapy for skull base adenoid cystic carcinoma. Arch Otolaryngol Head Neck Surg 2006; 132:1242–1249.
34 Linton OR, Moore MG, Brigance JS, et al: Proton therapy for head and neck adenoid cystic carcinoma: initial clinical outcomes. Head Neck 2015;37:117–124.
35 Owen D, Iqbal F, Pollock BE, et al: Long-term follow-up of stereotactic radiosurgery for head and neck malignancies. Head Neck 2015;37:1557–1562.
36 Lee N, Millender LE, Larson DA, et al: Gamma knife radiosurgery for recurrent salivary gland malignancies involving the base of skull. Head Neck 2003;25: 210–216.
37 Wang CC, Goodman M: Photon irradiation of unresectable carcinomas of salivary glands. Int J Radiat Oncol Biol Phys 1991;21:569–576.
38 Cerda T, Sun XS, Vignot S, et al: A rationale for chemoradiation (vs radiotherapy) in salivary gland cancers? On behalf of the REFCOR (French rare head and neck cancer network). Crit Rev Oncol Hematol 2014;91:142–158.
39 Pignon JP, le Maître A, Maillard E, et al: Meta-analysis of chemotherapy in head and neck cancer (MACH-NC): an update on 93 randomised trials and 17,346 patients. Radiother Oncol 2009;92:4–14.
40 Maruya S, Namba A, Matsubara A, et al: Salivary gland carcinoma treated with concomitant chemoradiation with intra-arterial cisplatin and docetaxel. Int J Clin Oncol 2006;11:403–406.
41 Pederson AW, Haraf DJ, Blair EA, et al: Chemoirradiation for recurrent salivary gland malignancies. Radiother Oncol 2010;95:308–311.
42 Samant S, van den Brekel MW, Kies MS, et al: Concurrent chemoradiation for adenoid cystic carcinoma of the head and neck. Head Neck 2012;34:1263–1268.
43 Airoldi M, Gabriele AM, Gabriele P, et al: Concomitant chemoradiotherapy followed by adjuvant chemotherapy in parotid gland undifferentiated carcinoma. Tumori 2001;87:14–17.

44 Katori H, Tsukuda M: Concurrent chemoradiotherapy with cyclophosphamide, pirarubicin, and cisplatin for patients with locally advanced salivary gland carcinoma. Acta Otolaryngol 2006;126:1309–1314.

45 Surakanti SG, Agulnik M: Salivary gland malignancies: the role for chemotherapy and molecular targeted agents. Semin Oncol 2008;35:309–319.

46 Ko HC, Gupta V, Mourad WF, et al: A contouring guide for head and neck cancers with perineural invasion. Pract Radiation Oncol 2014;4:e247–e258.

47 Lanning R, Young R, Barker C, et al: Cranial nerves; in Lee N, Riaz N, Lu J (eds): Target Volume Delineation for Conformal and Intensity-Modulated Radiation. New York, Springer, 2014, pp 167–203.

48 Lloyd S, Yu JB, Ross DA, et al: A prognostic index for predicting lymph node metastasis in minor salivary gland cancer. Int J Radiat Oncol Biol Phys 2010;76:169–175.

49 Chen AM, Garcia J, Lee NY, et al: Patterns of nodal relapse after surgery and postoperative radiation therapy for carcinomas of the major and minor salivary glands: what is the role of elective neck irradiation? Int J Radiat Oncol Biol Phys 2007;67:988–994.

50 Armstrong JG, Harrison LB, Thaler HT, et al: The indications for elective treatment of the neck in cancer of the major salivary glands. Cancer 1992;69:615–619.

Ana Kiess, MD, PhD
Department of Radiation Oncology and Molecular Radiation Sciences
Johns Hopkins University, 401 North Broadway, Suite 1440
Baltimore, MD 21231 (USA)
E-Mail akiess1@jhmi.edu

Management of Recurrent Malignant Salivary Neoplasms

Mazin Merdad[a] · Jeremy D. Richmon[a] · Harry Quon[b]

[a] Department of Otolaryngology – Head and Neck Surgery, Johns Hopkins University, and [b] Department of Radiation Oncology and Molecular Radiation Sciences, Department of Otolaryngology – Head and Neck Surgery, School of Medicine, Baltimore, Md., USA

Abstract

The management of malignant salivary gland neoplasms is based on a surgical paradigm, with intraoperative findings and pathology guiding the role of local-regional adjuvant therapy. Despite high rates of local control, local relapse can be a dominant pattern of recurrence, presenting therapeutic challenges. Although an optimal management approach has not been established, aggressive salvage surgery is favored given the morbidity associated with tumor progression at the skull base and the lack of significant response associated with other available treatment modalities. Postoperative radiotherapy has been demonstrated to be effective in the initial management of malignant salivary gland neoplasms and is generally favored for recurrent, surgically resectable tumors for the appropriate patient. © 2016 S. Karger AG, Basel

Introduction

Recurrent salivary gland tumors (SGTs) are best managed by preventing their occurrence in the first place. Recurrence can be avoided in a reasonable number of SGT cases with the proper utilization of preoperative investigations, including imaging and fine needle aspiration biopsy, such that the most appropriate surgical intervention is chosen. Intraoperatively, the extent of surgical resection for a primary malignant lesion will be dictated by size of a tumor, involvement of the facial nerve, and any lymph node metastasis. Postoperative radiotherapy has been shown to decrease the rate of loco-regional recurrences for high-risk parotid tumors (based on certain postoperative pathological features). Appropriate management of primary malignant SGTs is summarized in table 1 [1].

Data from the Dutch Head and Neck Oncology Cooperative Group has shown that the risk of recurrence for all malignant SGTs treated with surgery with or without radiotherapy was 17% at 5 years and 22% at 10 years, while the overall survival (OS) was 64% at 5 years and 50% at 10 years. The risk of regional recurrence was <10% regardless of lymph node status. Distant metastasis was detected in 28% of the sample at 5 years and in 33% at 10 years. Therefore, SGTs are more likely to recur locally or distally and often present at an advanced stage [2].

Table 1. Management of primary malignant parotid tumors [1]

Risk level	Management	Survival rate
Low risk − T1, N0 (any pathology) − Acinic cell, myoepithelial, epithelial-myoepithelial, low-grade mucoepidermoid carcinoma, polymorphous low-grade adenocarcinoma, basal cell adenocarcinoma	− Surgery: partial parotidectomy, possible neck dissection of level II − Radiation: typical indications include tumor spillage and perineural invasion, with irradiation to the primary operative bed	5 years = 90–96%
Moderate risk − T2 and/or N1 (any pathology) − Intermediate- or high-grade mucoepidermoid carcinoma, adenoid cystic carcinoma, carcinoma ex-pleomorphic	− Surgery: subtotal versus total parotidectomy sparing VII if possible; consider selective dissection of levels Ib, II, III (consider level V if ear canal or postauricular involvement) − Radiation: radiation to the primary site and positive neck (typically ipsilateral), prophylactic dose to uninvolved ipsilateral neck is controversial − Chemotherapy: consider if close/positive margins or tumor spillage	5 years = 70–90%
High risk − T3/4, N2/3 (any pathology) − Salivary duct carcinoma, adenocarcinoma (not otherwise specified), carcinoma ex-pleomorphic, undifferentiated carcinoma	− Surgery: total versus radical parotidectomy; selective neck dissection of levels Ib, II, III, V; temporal bone dissection ± facial nerve grafting versus static suspension; ± free tissue transfer if bony involvement − Radiation: radiation to the primary site and positive neck, prophylactic dose to the remaining uninvolved neck is favored − Chemotherapy: consider if close/positive margins or tumor spillage	5 years = 47% T3, 19% T4

Follow-Up of Primary Tumors

The most recent (2014) National Comprehensive Cancer Network guidelines for SGT management suggests scheduling follow-up every 1–3 months in the first year after treatment, every 2–6 months in the second year, every 4–8 months in years 3–5, and yearly thereafter [3]. Posttreatment imaging is recommended at 6 months. The risk of recurrence in malignant SGTs is close to 20%, and the majority will recur in the first 5 years after treatment. Nonetheless, the risk of recurrence after 5 years remains considerable. Long-term data from the University of California, San Francisco showed late recurrences in 13 and 18% of patients after 10 and 15 years of disease-free follow-up, respectively [4]. The risk was highest in adenoid cystic carcinoma (26%), followed by carcinoma ex-pleomorphic adenoma (25%) and mucoepidermoid carcinoma (17%), and recommendations for longer follow-up are reasonable for these tumors.

Management of Recurrent Salivary Gland Tumors

Signs and Symptoms

Clinical signs and symptoms suggestive of local recurrence are dependent on the anatomic site of the primary tumor. Depending on the tumor's location, recurrence may include facial swelling, skin tethering, pain, nerve paralysis or paresthesia, dysphagia, and trismus. Cranial nerve neuropathy, excluding the facial nerve, is suggestive of advanced deep recurrence. An understanding of the neck and skull base anatomy, and clinical vigilance are required for the early diagnosis of local relapses. Regional relapse can occur and typically presents as a nodal mass, while distant relapses are typically asymptomatic.

Imaging

Contrast-enhanced MRI is the imaging modality of choice given its superior soft tissue resolution and ability to detect perineural invasion [5]. CT scans are helpful in delineating the extent of bony involvement, if present. Imaging work-up must include a metastatic survey given the high predilection of SGTs to recur distantly [2]. The utility of positron emission tomography-CT remains questionable. To our knowledge, there are no studies comparing the sensitivity and specificity of positron emission tomography-CT (PET-CT) to MRI in detecting the extent of malignant SGTs.

Tumor Management

Recurrent SGTs are best managed by a multidisciplinary team, including members from radiation oncology, medical oncology, pathology, otolaryngology, and speech and language pathology. Plastic and reconstructive surgeons are an integral part of the team in cases with potential facial nerve sacrifice or large potential defects. In addition, the assistance of an otologist may be required for mastoid/temporal bone drilling in cases where the facial nerve's main trunk is involved with tumor and needs to be traced further proximally. Opinions from each team member are crucial for reaching an ideal treatment plan for each individual patient. New radiological images should be reviewed closely with the radiologist to identify the extent of disease and the probability of complete resection. Additionally, pathology slides from the initial surgery, if reported elsewhere, should be reviewed again. Up to 7% of head and neck tumors will be diagnosed differently after a second opinion pathology review at a high-volume head and neck center [6].

Surgery

Aggressive treatment is required in almost all cases of locoregionally recurrent SGTs. The mindset should be geared towards eradicating the tumor bulk even at the expense of causing some potential functional deficit. The skin, facial nerve, muscles of mastication, zygomatic bone, mandible, and maxillary bone may require some form of re section. It must be noted that with proper treatment, including adjuvant therapy, a 5-year OS of 66% and a 5-year disease-free survival of 64% can be achieved [7].

Regional neck lymph nodes must be addressed if not dealt with during the primary surgery. The overall risk of lymph node involvement in parotid, submandibular gland, and minor salivary gland carcinomas is 25, 41, and 10%, respectively [2]. Except for localized low-grade parotid tumor recurrences, elective neck dissection of levels Ib, II, and III is warranted. Excision of levels IV and V might be warranted in stages rIII and rIV and in cases with external auditory canal or post auricular involvement. In recurrent submandibular gland tumors, excision of levels I, II, and III (+/– IV in stages rIII and rIV) is recommended. Neck dissection of levels I, II and III should also be considered in recurrent minor salivary gland carcinomas that involve the tongue, floor of the mouth, or soft palate.

A form of facial nerve monitoring, such as electromyography or direct observation of facial twitching with stimulation, is essential in recurrent parotid and submandibular gland surgery. Small and superficial tumors with no gross nerve involvement may be excised with the nerve left intact. However, this is more often the exception, and recurrent carcinomas very frequently require some degree of facial nerve sacrifice. Sacrificing the nerve is required when the face is paralyzed preoperatively or when the nerve is intraoperatively found to be grossly involved with a tumor. Occasionally, using sharp meticulous dissection may permit peeling all gross tumor off of the nerve without any major injury to the nerve. If a tumor's extension warrants facial nerve sacrifice, the free edges of the nerve should be sent for intraoperative frozen section to confirm negative margins and complete tumor resection.

The extent of facial nerve rehabilitation is uniquely tailored to each individual case. The facial nerve is preferably reconstructed at the time of surgery to restore neuronal input to the facial mimetic muscles and maintain their tone. Options include direct reanastomosis, cable grafting, nerve crossover, and cross-facial nerve grafting. If nerve reanastomosis is not possible, static reconstructive options, such as upper eyelid gold/platinum weight implants and elevation of the oral commissure can be utilized. Delayed dynamic reconstructive options include temporalis, masseter, and gracilis neuromuscular unit implantation [8].

The defects created after the excision of recurrent tumors can vary based on their local extension. Aggressive recurrent parotid tumors will often necessitate cortical mastoidectomy, temporal bone resection, and possibly excision of parts of the zygomatic bone or mandibular bone. A cortical mastoidectomy may be useful in circumstances where the facial nerve needs to be proximally exposed or if the pinna or any of the ear chambers are involved. Mastoid surgery can help achieve negative margins in cases with proximal nerve involvement. In a retrospective review from MD Anderson Cancer Center regarding patients with advanced or recurrent parotid tumors, 51% of the patients were found to have perineural invasion of the main nerve trunk when a mastoidectomy was performed. Nonetheless, achieving negative margins was possible in up to 78% of the cohort with the aid of mastoidectomy [9]. In cases where there is limited external auditory canal involvement, sleeve resection with cortical mastoidectomy might suffice; however, more extensive external canal involvement will require a formal lateral temporal bone resection in order to achieve negative margins. Similarly, advanced recurrent submandibular gland and minor SGTs might require marginal or segmental mandibulectomy and low or infrastructure maxillectomy, respectively. Free fibular or scapular bone flaps are commonly used options used to fill in segmental bony defects [10]. For augmenting large parotid soft tissue defects with or without skin loss, anterolateral thigh free flaps have been shown to reasonably recreate the contralateral facial contour [11].

Radiation
Role of Radiation for Adjuvant Indications
Adjuvant external beam radiation offers a survival advantage when used in the treatment of malignant SGTs. A systematic review of parotid cancer publications by Jeannon et al. [12] found a significant OS advantage in patients treated with adjuvant radiotherapy (hazard ratio = 2.5, 95% CI = 1.5–4.7). Similarly, a population-based study from the United States using the Surveillance, Epidemiology, and End Results Program database [13] revealed a significant survival advantage in submandibular gland carcinoma patients treated with postoperative radiotherapy (hazard ratio = 1.7, 95% CI = 1.09–2.36). The Dutch Oncology Head and Neck Cooperative Group [14] observed improved local and regional control rates with postoperative radiotherapy for certain pathologic indications, including T3–4 tumors, incomplete or close resection, bone invasion, perineural invasion and positive persistent nodal disease status. Long-term follow-up data for malignant minor SGTs treated with surgery followed by radiation has shown local control rates of 90% at 5 years and 88% at 10 years [15].

In addition to decisions regarding adjuvant radiation to a postoperative primary tumor site, decisions must be made regarding whether elective nodal irradiation (ENI) should be considered, especially when the pathologic status of the neck is unknown. In a retrospective analysis, Chen et al.

[16] demonstrated improved regional control rates with ENI. The greatest crude rates of neck relapse were found in patients not receiving ENI and were dependent on tumor stage and histology, especially high-risk histology, such as adenocarcinoma, mucoepidermoid carcinoma (especially when scored as high grade), undifferentiated carcinoma and squamous carcinoma. There were no neck relapses observed in patients with acinic cell carcinoma or adenoid cystic carcinoma with or without ENI. Thus, the primary site indications for ENI include advanced tumor stage and high-risk histology. Although Chen et al. [16] did not identify an increased risk of nodal relapse with primary tumor involvement in lymphatic-rich regions, such as the oral cavity, this remains a prudent consideration for each individual patient. Similar conclusions were observed by Terhaard et al. [14] in their review of 538 patients in the Dutch Head and Neck Oncology Cooperative Group, offering confidence in the validity of these indications. Terhaard's study recommended the use of at least 60 Gy.

Role of Radiation for Unresectable Indications
There was considerable interest in neutron radiotherapy in the early 1990s after the Radiation Therapy Oncology Group (RTOG) in the United States and the Medical Research Council in Great Britain published their data from a randomized controlled trial comparing neutron radiotherapy against conventional photon and/or electron radiotherapy for the treatment of unresectable malignant SGTs. Such interest has been centered on the potential favorable radiobiological effects of using neutron radiation to overcome radiation resistance. While neutron radiation-treated patients did have lower rates of locoregional recurrence, their OS was not found to be significantly different. Further, the rates for severe treatment complications were higher in the neutron radiation-treated group [17]. The University of Washington in Seattle has published extensively on the use of neutron therapy in the treatment of advanced and unresectable submandibular gland tumors with good overall results, although these reports have also highlighted the adverse effects in normal tissue associated with the higher radiobiologic effects of neutron particle radiation [18]. Nonetheless, the use of neutron therapy remains limited given the paucity of neutron therapy centers and the relatively excellent outcomes achieved using modern photon therapy delivery techniques [19].

For unresectable malignant SGTs, the Dutch Head and Neck Oncology Cooperative Group observed a clear dose-response relationship when treatment involved radiotherapy alone. A 5-year local control rate of 50% was observed in patients treated with a dose range of 66–70 Gy [14]. Whether further improvements in local control rates may be achieved with the use of altered fractionation schedules, such as accelerated hyperfractionated schedules, is unclear; improved rates have been reported by Wang and Goodman [20], although with limited follow-up.

Role of Radiation for Recurrent Indications
Data regarding the reirradiation of recurrent SGTs are limited. Pederson et al. [21] retrospectively reviewed their experiences with 14 recurrent resectable SGT patients treated with postoperative chemo-reirradiation. The median time to locoregional recurrence was 13 months. The median initial radiation dose was 60 Gy. The median interval to reirradiation was 48 months, and the median second radiation dose was 66 Gy. With a median follow-up duration of 18 months (range 2–125 months), postoperative chemo-reirradiation was able to achieve locoregional control rates of 72 and 52% at 1 and 3 years, respectively. The 5-year OS was 27% [21]. Although only acute toxicities were reported in 12 of the 14 patients, it is noteworthy that 6 of the 12 had polyethylene glycol tubes in place at the time of last follow-up or death, indicating the potential for adverse normal tissue following salvage surgery and postoperative chemo-reirradiation. The placement of unirradiated tissue in the reconstruction phase of any salvage surgery as opposed to primary closure is

strongly recommended as a strategy to minimize adverse normal tissue effects from reirradiation. Proper selection of patients with minimal late radiation effects from the first course of radiation is also an important consideration for minimizing the risk of reirradiation [22].

Stereotactic radiotherapy (SRT) is another delivery modality using conventional photon radiation that has been used for treating recurrent SGTs previously treated with external beam radiation. SRT offers the potential advantage of delivering larger radiotherapy doses per treatment fraction, which offers a greater probability not only for tumor cytotoxicity but also for potentially vascular ablative effects. To overcome the risk of injuring surrounding normal tissues, the delivery technique requires greater patient immobilization and machine hardware to ensure daily precision. Karam et al. treated a series of 18 patients with recurrent SGTs using CyberKnife® SRT. Eight patients from the series had revision surgery and 7 had chemotherapy. The median radiation dose was 30 Gy in 5 fractions. With a median follow-up of 12 months, the 2-year local-regional control and OS rates were 53 and 39%, respectively. Severe late toxicity in the form of soft-tissue necrosis occurred in 4 patients (22%) and was correlated with radiation dose [23].

Chemotherapy

It is important to note that to date there are no phase III trials studying the efficacy of postoperative chemotherapy in recurrent/metastatic malignant SGTs. Data from retrospective studies have not been able to demonstrate any significant improvement in OS [24–27].

Chemotherapy is utilized in the setting of recurrent/metastatic SGTs mainly for palliation or as an adjunct to radiation after surgical excision of high-risk tumors. Conventional chemotherapy with platinum-based single agents or in combination with doxorubicin or cyclophosphamide is the most widely used regimen for recurrent and metastatic malignant SGTs. Recurrent adenoid cystic carcinomas treated with combination chemotherapy have shown a response rate between 20 and 25%, lasting anywhere between 2 and 10 months [24]. Combination chemotherapy patients had more toxic side effects than single chemotherapy patients.

In the setting of postoperative high-risk tumors, small retrospective reports have suggested a possible benefit from the addition of platinum-based single agent chemotherapy to radiation. The associated toxicities seem to be well tolerated, especially by younger age groups [25, 26]. Radiation Therapy Oncology Group study number 1008 is a randomized phase II trial comparing adjuvant radiation to concurrent chemoradiation treatment in resected, high-risk, malignant SGTs. The trial closed to accrual in March 2015 after the enrollment of 124 patients [27].

Molecular therapies targeting epidermal growth factor (e.g., cetuximab and gefitinib), human epidermal growth factor receptor 2 (e.g., trastuzumab and lapatinib), c-KIT (e.g., imatinib), and vascular endothelial growth factor (e.g., axitinib) have been recently investigated in the setting of recurrent/metastatic malignant SGTs [28, 29]. Unfortunately, none of the trials using molecular therapies were able to demonstrate any objective response. There has been recent interest in an *MYB-NFIB* fusion oncogene resulting from gene translocation, which has been found in one third of acinic cell carcinoma tumors. Different targeted therapies against the translocated gene are under investigation [30].

References

1 Gillespie MB, Albergotti WG, Eisele DW: Recurrent salivary gland cancer. Curr Treat Options Oncol 2012;13:58–70.
2 Terhaard CH, Lubsen H, van der Tweel I, et al: Salivary gland carcinoma: independent prognostic factors for locoregional control, distant metastases, and overall survival: results of the Dutch head and neck oncology cooperative group. Head Neck 2004;26:681–692; discussion 692–693.

3 Chen AM, Garcia J, Granchi PJ, et al: Late recurrence from salivary gland cancer: when does 'cure' mean cure? Cancer 2008;112:340–344.
4 http://www.nccn.org/professionals/physician_gls/f_guidelines.asphttp://www.nccn.org/professionals/physician_gls/f_guidelines.asp.
5 Shah GV: MR imaging of salivary glands. Neuroimaging Clin N Am 2004; 14:777–808.
6 Westra WH, Kronz JD, Eisele DW: The impact of second opinion surgical pathology on the practice of head and neck surgery: a decade experience at a large referral hospital. Head Neck 2002;24: 684–693.
7 Kobayashi K, Nakao K, Yoshida M, et al. Recurrent cancer of the parotid gland: how well does salvage surgery work for locoregional failure? ORL J Otorhinolaryngol Relat Spec 2009;71:239–243.
8 Boahene K: Reanimating the paralyzed face. F1000Prime Rep 2013;5:49.
9 Gidley PW, Thompson CR, Roberts DB, et al: The results of temporal bone surgery for advanced or recurrent tumors of the parotid gland. Laryngoscope 2011;121:1702–1707.
10 Clark JR, Vesely M, Gilbert R: Scapular angle osteomyogenous flap in postmaxillectomy reconstruction: defect, reconstruction, shoulder function, and harvest technique. Head Neck 2008;30: 10–20.
11 Cannady SB, Seth R, Fritz MA, et al: Total parotidectomy defect reconstruction using the buried free flap. Otolaryngol Head Neck Surg 2010;143:637–643.
12 Jeannon JP, Calman F, Gleeson M, et al: Management of advanced parotid cancer. A systematic review. Eur J Surg Oncol 2009;35:908–915.
13 Bhattacharyya N: Survival and prognosis for cancer of the submandibular gland. J Oral Maxillofac Surg 2004;62: 427–430.
14 Terhaard CH, Labsen H, Rasch CR, et al: The role of radiotherapy in the treatment of malignant salivary gland tumours. Int J Radiat Oncol Bio Phys 2005;61:103–111.
15 Zeidan YH, Shultz DB, Murphy JD, et al: Long-term outcomes of surgery followed by radiation therapy for minor salivary gland carcinomas. Laryngoscope 2013;123:2675–2680.
16 Chen A, Garcia J, Lee MK, et al: Patterns of nodal relapse after surgery and postoperative radiotherapy for carcinoma of the major and minor salivary glands: What is the role of elective neck dissection? Int J Radiat Oncol Bio Phys 2007; 67:988–994.
17 Laramore GE, Krall JM, Griffin TW, et al: Neutron versus photon irradiation for unresectable salivary gland tumors: final report of an RTOG-MRC randomized clinical trial. Radiation Therapy Oncology Group. Medical Research Council. Int J Radiat Oncol Biol Phys 1993;27:235–240.
18 Laramore GE: Role of particle radiotherapy in the management of head and neck cancer. Curr Opin Oncol 2009;21: 224–231.
19 Spratt DE, Salgado LR, Riaz N, et al: Results of photon radiotherapy for unresectable salivary gland tumors: is neutron radiotherapy's local control superior? Radiol Oncol 2014;48:56–61.
20 Wang CC, Goodman M: Photon irradiation of unresected carcinomas of the salivary gland. Int J Radiat Oncol Biol Phys 1991;21:569–576.
21 Pederson AW, Haraf DJ, Blair EA, et al: Chemoreirradiation for recurrent salivary gland malignancies. Radiother Oncol 2010;95:308–311.
22 Stevens KR, Britsch A, Moss WT: High dose reirradiation of head and neck cancer with curative intent. Int J Radiat Oncol Biol Phys 1994;29:687–698.
23 Karam SD, Snider JW, Wang H, et al: Reirradiation of recurrent salivary gland malignancies with fractionated stereotactic body radiation therapy. J Radiat Oncol 2012;1:147–153.
24 Laurie SA, Ho AL, Fury MG, et al: Systemic therapy in the management of metastatic or locally recurrent adenoid cystic carcinoma of the salivary glands: a systematic review. Lancet Oncol 2011; 12:815–824.
25 Tanvetyanon T, Qin D, Padhya T, et al: Outcomes of postoperative concurrent chemoradiotherapy for locally advanced major salivary gland carcinoma. Arch Otolaryngol Head Neck Surg 2009;135: 687–692.
26 Schoenfeld JD, Sher DJ, Norris CM Jr, et al: Salivary gland tumors treated with adjuvant intensity-modulated radiotherapy with or without concurrent chemotherapy. Int J Radiat Oncol Biol Phys 2012;82:308–314.
27 www.rtog.org/ClinicalTrials/ProtocolTable/StudyDetails.aspx?study=1008.
28 Surakanti SG, Agulnik M: Salivary gland malignancies: the role for chemotherapy and molecular targeted agents. Semin Oncol 2008;35:309–319.
29 Andry G, Hamoir M, Locati LD, et al: Management of salivary gland tumors. Expert Rev Anticancer Ther 2012;12: 1161–1168.
30 Stenman G, Andersson MK, Andrén Y: New tricks from an old oncogene: gene fusion and copy number alterations of MYB in human cancer. Cell Cycle 2010; 9:2986–2995.

Harry Quon, MD, MS(CRM), Co-Director, Multidisciplinary Head and Neck Program, Associate Professor
Department of Radiation Oncology and Molecular Radiation Sciences
Department of Otorhinolaryngology – Head and Neck Surgery, Department of Oncology
Sidney Kimmel Comprehensive Cancer Center
401 North Broadway-Suite 1440, Baltimore, MD 21231 (USA)
E-Mail hquon2@jhmi.edu

Salivary Gland Neoplasms in Children and Adolescents

Patrick J. Bradley[a] · David W. Eisele[b]

[a]School of Medicine, The University of Nottingham, Nottingham University Hospitals, Queens Medical Centre Campus, Nottingham, UK; [b]Department of Otolaryngology – Head and Neck Surgery, Johns Hopkins University School of Medicine, Baltimore, Md., USA

Abstract

Salivary gland neoplasms (SGNs) in children are uncommon. Epithelial SGNs (ESGNs) comprise the majority (95%), with the remaining being mesenchymal SGNs (MeSGNs). Pleomorphic adenoma is the most frequently encountered benign neoplasm, mucoepidermoid carcinoma is the most frequent malignant ESGN, and rhabdomyosarcoma is the most frequent malignant MeSGN. ESGN presents in the second decade, whereas MeSGN presents in the first and second decades. Swelling without pain or neurological signs is the main presentation of both benign and malignant neoplasms. Making an accurate preoperative histological diagnosis is important, so a needle biopsy or a perioperative frozen section is useful when there is doubt about the disease status of the patient; the excised tumour margin is also important. Surgical excision should aim to achieve clear margin excision in benign and malignant ESGNs, minimising the need for adjuvant radiotherapy and maximising the long-term likelihood of patient cure. Benign ESGNs are uncommon, and excision is curative, whereas malignant ESGN and MeSGN should be managed by a multidisciplinary paediatric oncology team.

© 2016 S. Karger AG, Basel

Introduction

Salivary gland neoplasms (SGNs) are uncommon in children. They comprise <10% of all paediatric head and neck tumours and account for only 5% of all salivary gland malignancies. The majority (>95%) are epithelial, with a minor group being mesenchymal in origin. When a solitary SGN is present, the chance of malignancy is greater in a child than in an adult [1–3]. The concept that paediatric SGNs are more likely malignant than adult SGNs has been challenged by suggestions that previous data have been 'selected material' and by reporting of a nationwide survey from Denmark over a 16-year period (1990–2005) [4]. The survey identified 61 patients, in whom 85% of the tumours were benign, and the parotid gland was involved in the 9 cases of malignancy. Most paediatric epithelial neoplasms occur in children >10 years of age (median age 11–14 years). If malignant, the majority are classified low or intermediate grade. Malignant tumours in children <10 years tend to be high grade and associated with a poor prognosis.

Table 1. Differential diagnosis of a non-inflammatory salivary gland mass

Neoplastic type	Benign	Malignant
Epithelial or parenchymal origin	Pleomorphic adenoma	Mucoepidermoid carcinoma Acinic cell carcinoma Adenoid cystic carcinoma Adenocarcinoma
Mesenchymal or interstitial origin	Vascular malformation Neurofibroma Schwannoma	Rhabdomyosarcoma Liposarcoma Aggressive fibromatosis
Other 'tumour-like' lesions	Pilomatrixoma Reparative granuloma Reactive lymph node Sinus histiocytosis Granular cell tumour Benign simple cyst First branchial cleft cyst	Lymphoma

The parotid gland is involved in >85% of all reported paediatric SGNs [1–4]. Minor SGNs do occur, with the majority located in the oral cavity; other sites have also been reported, usually as case reports, including the pharynx, larynx and bronchus [5].

The most common benign neoplasm affecting the paediatric major salivary glands are vascular malformations, namely, haemangiomas and lymphangiomas [6, 7]. Haemangiomas are the most common vascular tumour of childhood, which, after a growth phase, undergo spontaneous involution, a process lasting months or occasionally years. Haemangiomas are reported to account for almost 60% of tumours in children, of which 80% occur in the parotid gland, with involvement of the submandibular region in 18% of cases, while 2% are associated with the minor salivary glands [8].

Mesenchymal SGNs (MeSGNs) are very rarely reported in children. Benign MeSGNs, such as neurofibroma and schwannoma (table 1), have been reported, mainly as case reports, in the major salivary glands [9]. Malignant MeSGNs have been reported, with rhabdomyosarcoma (RMS) being the most common soft tissue malignancy seen, with a median age of 7 years (range 0–19 years) [6, 10]. RMS of the embryonal and undifferentiated types presents most commonly at <5 years of age, with all types of RMS, including the alveolar type, presenting in the older child. About 40% of all RMSs arise in the head and neck, with the 'parotid region' being involved by direct invasion. The risk of RMS is higher in children with Li-Fraumeni syndrome, neurofibromatosis type I, Beckwith-Wiedemann syndrome, nevoid basal cell carcinoma (Gorlin) syndrome, and Rubinstein-Taybi syndrome. Other malignant MeSGNs have been reported in children [6, 11, 12], including liposarcoma and fibrosarcoma, among others. All children diagnosed with a head and neck sarcoma require multimodality therapy to maximise local tumour control. A decision as to the role and place of surgery and chemoradiotherapy in the treatment strategy needs to be discussed and planned in an agreed sequence by a multidisciplinary paediatric oncology team. RMS is usually curable in children with localised disease, with >70% surviving 5 years [10].

Presentation

The most common presenting sign or symptom in the major salivary glands is an obvious mass (in 100% of patients) that has developed or is enlarging. The average duration of a mass before clinical presentation is 8–12 months. Pain is notably absent as a dominant symptom in children presenting with an SGN, whether benign or malignant [6, 9, 13]. The incidence of facial paresis approaches <4%, and evidence of cervical lymphatic metastasis is reported in 3.5% of patients at presentation [6]. For minor salivary glands, benign and malignant neoplasms most commonly present in the oral cavity as a 'lump' or submucosal nodule. The majority of neoplasms are non-ulcerated, with some firm to palpation, with pink or flesh-coloured surfaces, and others are described as being fluctuant, with colour alterations ranging from a light blue hue to purple [14].

Differential Diagnosis and Investigation

A 'discrete' parotid mass in children is uncommon and can represent various pathological diagnoses. Two large series of children and adolescents [9, 13] at tertiary teaching paediatric hospitals in the USA have reported the spectrum of diagnoses (table 1). In the infant/younger child, vascular malformations are the most common diagnosis. The diagnosis of haemangioma is based on inspection, history, and physical examination, with lesions having varying degrees of compressibility to the limits of their underlying fibro-fatty architecture. If lesions are more complex or if surgical intervention is required, additional investigation may be helpful in determining the extent of involvement. In such cases, a variety of non-invasive imaging modalities are extremely helpful in verifying the diagnosis. It is important to identify lesions caused by infection and inflammation (sialadenitis +/– abscess formation), which become more frequent with advancing age. It is important to separate the solid lesions from the semisolid or cystic lesions [6, 9, 13].

The use of ultrasound (colour flow Doppler) is very useful for diagnosing vascular malformations as well as for identifying cystic and semisolid masses in the major salivary glands. The use of computed tomography and magnetic resonance imaging helps to delineate the extent and depth of surrounding tissue involvement. However, a recent report suggested that computed tomography is not helpful for distinguishing benign and malignant disease but that magnetic resonance imaging features, such as T2 hypointensity, restricted diffusion, ill-defined borders, and focal necrosis, although not specific, should raise concerns about malignancy [15]. Occasionally, the use of fine-needle aspiration cytology or fine-needle core biopsy may be required to avoid major surgery in situations such a mass due to lymphoma [16, 17]. The clinical utility of needle biopsy remains debatable, however, especially in younger patients, because of the need for an anaesthetic. If the lesion or mass is considered 'small enough', can be excised without significant morbidity, and involves the parotid or submandibular glands, then a plan to proceed to excision biopsy after a course of an antibiotic and thorough investigation may be reasonable [18]. Obviously, a frank and honest discussion with the parents or guardians about the likely diagnosis, the purpose of the procedure, and the likely outcome is important. Employing a perioperative frozen section, if available, may help to reassure the surgeon of 'patient-side tumour-free margins' after excision surgery.

Epithelial Salivary Gland Neoplasms (ESGN)

Benign

Pleomorphic adenoma (PA) accounts for >50% of all epithelial tumours in children [19]. Composite analysis of several studies suggests that most (62%; range 56–77%) PAs occur in the parotid gland and submandibular gland (26%; range 11–40%) and in the minor salivary glands (12%; range 0–21%). The sex distribution is 1.4 females:1 male, and presentation is commonly between the ages of 8 and 18 years [6, 19–21]. Typically, the tumour presents as a slow-growing, painless, firm mass present for an average duration of >12 months. The tumour is a single-site origin and is well-circumscribed and lobulated, with a smooth surface, when small and then becomes bosselated with increasing size. The treatment of choice is complete surgical resection. Recurrences are to be expected if the tumour is enucleated, but recurrence may not be clinically evident for many years.

PA is the most frequent ESGN diagnosed among minor SGNs presenting between the ages of 19 months and 18 years (mean age 12 years). There is a slightly higher female preponderance, at 2.8:1. The majority are located in the hard and/or soft palate, followed by the buccal mucosa, upper lip and tongue [22]. Benign minor salivary neoplasms are reported at all ages in children, but most occur in the second decade [6, 19–27]. The other described benign epithelial salivary gland tumours have been encountered in the major salivary glands and should be treated similarly to PAs [21].

Malignant
Malignancies of the salivary glands in children and adolescents are rare, with an estimated annual incidence of 0.08 per 100,000 (for the parotid and submandibular glands) [17]. However, the Danish nationwide survey reported incidences of 0.12 and 0.53 per 100,000 children for malignant and benign ESGNs, respectively [4]. Most malignant salivary gland tumours present in the older child, with an average age of 13.5 years at diagnosis, whereas benign neoplasms present at a slightly older age, or 15 years, with an equal sex distribution [6]. The most frequent site reported for malignant salivary tumours is the parotid gland (82%), followed by the submandibular gland (7%), minor salivary glands (11%) and sublingual gland (<1%) [6]. Mucoepidermoid carcinoma (MEC) is the most common ESGN malignancy (60%), followed by acinic cell carcinoma (11%) and adenocarcinoma (not otherwise specified) (10%), while others, such as adenoid cystic carcinoma (9%), carcinoma ex-PA (1%) and squamous cell carcinoma (<1%) are considered rare [6].

Malignant tumours of the minor salivary glands can be difficult to diagnose because of their site dependent presentation. The majority are located in the oral cavity and histologically are MEC. Other sites reported include the nasopharynx [28], larynx (subglottic) [29], and piriform fossa [30].

Staging of Salivary Gland Malignancies

The TNM staging system (used for adults) is used for children for the major salivary glands, whereas for minor salivary gland malignancies, the staging used is the system for squamous cell carcinoma disease, which is tumour site dependent [discussed further in the chapter by Bradley, this vol. pp. 9–16]. Children are more often diagnosed at an early stage with malignant disease, which is low- or intermediate-grade disease compared to disease in adults [5, 20].

Treatment

The treatment of choice for SGNs is complete removal of the tumour mass, with a clear surgical margin.

Parotid Gland
Parotidectomy consists of identification of the facial nerve and its preservation, followed by resection of a surrounding portion of the salivary tissue in which the tumour is located [23]. Superficial or lateral parotidectomy for a laterally lo-

cated tumour mass is the typical procedure. Total parotidectomy may be required for deep lobe or parapharyngeal tumour mass lesions [27] or when the tumour mass is suspected or confirmed to be high grade or to have aggressive malignant potential, such as a mass presenting with a facial nerve palsy, multiple intraparotid masses, or a mass coupled with the presence of cervical metastasis [2, 5, 6, 16, 20, 31–33]. Resection of the facial nerve is currently recommended when there is gross anatomic or histological evidence of nerve invasion at the time of surgery. Should the nerve trunk, nerve division or a named branch be resected, immediate nerve repair by means of a primary anastomosis or a free nerve graft is advocated [2, 33].

Submandibular Gland
Removal of the entire gland that contains the tumour mass is recommended. It may, on occasion, be possible to remove the gland with an adequate margin when a benign tumour is present, leaving a functioning portion of the normal gland [34]. However, when a diagnosis of a malignant process is suspected due to local tissue involvement observed radiologically and proven by fine-needle aspiration cytology/fine-needle core biopsy or during surgery, then a local resection of level Ib (selective neck dissection), extended as indicated, should be performed to ensure that the primary and local soft tissues are removed 'en bloc'.

Sublingual and Minor Salivary Glands
Primary malignant tumours of the sublingual gland in children are exceedingly rare, with the majority being malignant [21, 31]. Minor SGNs are more frequently benign than malignant, with the majority reported to arise in the oral cavity [16, 31]. MEC is the most frequent malignant pathology among sublingual and intraoral malignant minor salivary gland tumours. The majority are classified as low- to intermediate-grade tumours. MEC has a female predilection and is decidedly uncommon in the first decade of life.

MEC most frequently occurs in the hard or soft palate or both, followed by the buccal mucosa and upper lip. Complete surgical excision is the treatment of choice and may include partial mandibulectomy if the periosteum has been invaded [35]. Such surgical principles result in a recurrence rate of <10% [14].

The Neck
Therapeutic neck dissection should be performed when palpable or when imaging studies demonstrate cervical lymph node involvement. Elective neck dissection may be considered for high-grade and advanced-stage tumours. Postoperative radiation therapy may be avoided in pN0 necks [31, 33, 35].

Postoperative Radiotherapy

Postoperative radiotherapy may be indicated for patients with adverse histological features or recurrent disease [6, 31, 33, 36]. Radiation therapy for the neck should be considered on a case-by-case basis for pN+ necks. Extracapsular spread or involvement of more than 3 nodes are solid indications. In other cases, irradiation volumes should be limited to the primary site for high-grade tumours, given the potential long-term side effects of radiation in children [6, 33, 35, 36].

Recurrence

The recurrence rate of benign salivary neoplasms in children is >20%, which is explained by the greater risk of rupture during surgery by inexperienced surgeons, as a consequence of the reduced size of the anatomical structures, or by misdiagnosis and performance of very conservative surgery. Revision surgery is associated with an increased risk of facial nerve injury and the potential need for excision of facial skin or other structures [5].

Recurrence of malignant epithelial neoplasms is more often associated with adenoid cystic carcinoma than with MEC. Recurrences are dependent on the tumour stage and grade and the extent of surgery. Revision surgery is associated with significant morbidity, and the majority, and more so the high-grade neoplasms, will not achieve a long-term benefit [37].

Outcome

The survival of paediatric patients with major salivary gland carcinomas is generally favourable. Adverse outcomes are best predicted by the tumour grade, the margin status, and neural involvement. Radiation therapy is beneficial for loco-regional control of disease, with acceptable long-term treatment sequelae and without significant risk of developing second primary tumours. The overall survival is 70–90% at 5 years, and 26% develop a recurrence [5, 6, 20, 22, 31, 32, 35, 37]. Paediatric salivary gland cancer patients are at a 3% risk of developing a second cancer within 20 years after completion of treatment of the primary malignancy. The expected risk of a subsequent neoplasm increases by 3- to 6-fold the expected risk, and previous radiotherapy is possibly also implicated [37].

References

1 Muenscher A, Diegel T, Jaehne M, et al: Benign and malignant gland diseases in children; a retrospective study of 549 cases from the salivary gland registry, Hamburg. Auris Nasus Larynx 2009;36: 326–331.
2 Yoshida EJ, Garcia J, Eisele DW, et al: Salivary gland malignancies in children. Int J Pediat Otorhinolaryngol 2014;78: 174–178.
3 da Cruz Perez DE, Pires FR, Alves FA, et al: Salivary gland tumors in children and adolescents: a clinicopathologic and immunohistochemical study of fifty-three cases. Int J Pediatr Otorhinolaryngol 2004;68:895–902.
4 Stevens E, Andreasen S, Bjørndal K, et al: Tumors of the parotid and not relatively more often malignant in children than in adults. Int J Pediatric Otorhinolaryngol 2015;79:1192–1195
5 Bradley P, McClelland L, Metha D: Paediatric salivary gland epithelial neoplasms. ORL 2007;69:137–145.
6 Lennon P, Cunningham MJ: Salivary gland tumors; in Rahbar R (ed): Pediatric Head and Neck Tumors: A–Z Guide to Presentation and Multimodality Management. Boston, Springer Verlag, 2015, pp 311–327.
7 Dasgupta R, Fishman SJ: International Society for the Study of Vascular Anomalies (ISSVA) classification. Sem Ped Surg 2014;23:158–161.
8 Iro H, Zenk J: Salivary gland diseases in children. GMS Curr Top Otorhinolaryngol Head Neck Surg 2014;13:Doc06.
9 Bentz BG, Hughes A, Ludermann JP, et al: Masses of the salivary region in children. Arch Otolaryngol Head Neck Surg 2000;126:1435–1439.
10 Cockerill CC, Daram S, El-Naggar AK, et al: Primary sarcomas of the salivary glands: case series and literature review. Head Neck 2013;35:1551–1557.
11 Takahama A, Leon JE, de Almeida OP, et al: Non-lymphoid mesenchymal tumours of the parotid gland. Oral Oncol 2008;44:970–974.
12 Cho K-J, Ro JY, Choi J, et al: Mesenchymal neoplasms of the major salivary glands: clinicopathological features of 18 cases. Eur Arch Otorhinolaryngol 2008; 265 Suppl 1:S47–S56.
13 Orvidas LJ, Kasperbauer JL, Lewis JE, et al: Pediatric parotid masses. Arch Otolaryngol Head Neck Surg 2000;126:177–184.
14 Ritwik P, Cordell KG, Brannon RB: Minor salivary gland mucoepidermoid carcinoma in children and adolescents; a case series and review of literature. J Med Case Rep 2012;6:182.
15 Mamlouk MD, Rosbe KW, Glastonbury CM: Paediatric parotid neoplasms: a 10 year retrospective imaging and pathology review of these rare tumours. Clin Radiol 2015;70: 270–277.
16 Jaryszak EM, Shah RK, Bauman NM, et al: Unexpected pathologies in pediatric parotid lesions: management paradigms revisited. Int J Ped Otrhinolaryngol 2011;75:558–563.
17 Zielinski R, Kobos J, Zakrzewska A: Parotid gland tumors in children – pre- and post-operative diagnostic difficulties. Pol J Pathol 2014;65:130–134.
18 Jablenska L, Trinidade A, Merabnagri V, et al: Salivary gland pathology in the paediatric population: implications for management and presentation of a rare case. J Laryngol Otol 2014;128:104–106.
19 Fu H, Wang J, Wang L, et al: Pleomorphic adenoma of the salivary glands in children and adolescents. J Pediatric Surg 2012;47:713–719.

20 Sultan I, Rodriguez-Galindo C, Al-Sharabati C, et al: Salivary gland carcinomas in children and adolescents; a population-based study, with comparison to adult cases. Head Neck 2011;33:1476–1491.
21 Fang Q-G, Shi S, Li Z-N, et al: Epithelial salivary gland tumors in children; a twenty-five year experience of 122 patients. Int J Ped Otorhinolaryngol 2013; 77:1252–1254.
22 Ritwik P, Brannon RB: A clinical analysis of nine new pediatric and adolescent cases of benign salivary gland neoplasms and a review of the literature. J Med Case Rep 2012;6:287.
23 Rodriguez KH, Vargas S, Robson C, et al: Pleomorphic adenoma of the parotid gland in children. Int J Pediatr Otorhinolaryngol 2007;71:1717–1723.
24 Yu GY, Li ZL, Ma DQ: Diagnosis and treatment of epithelial salivary gland tumours in children and adolescents. Br J Oral Maxillofac Surg 2002;40:389–392.
25 Jorge J, Pires FR, Alves DE, et al: Juvenile intraoral pleomorphic adenoma: a report of five cases and review of the literature. Int J Oral Maxillofac Surg 2002;31:273–275.
26 Laikui L, Hongwei Li, Hongbing J, et al: Epithelial salivary gland tumors in children and adolescents in west China population: a clinicopathologic study of 79 cases. J Oral Pathol Med 2008;37:201–205.
27 Starek I, Mihal V, Novak Z, et al: Pediatric tumous of the parapharyngeal space; three case reports and a review of the literature. Int J Pediat Otorhinolaryngol 2004;68:601–606.
28 Re M, Pasquini E: Nasopharyngeal mucoepidermoid carcinoma in children Int J Pediat Otorhinolaryngol 2013;77:565–569.
29 Javadi M, Bafrouee FM, Mohseni M, et al: Laryngeal adenoid cystic carcinoma in a child. Ear Nose Throat J 2002;81:34–35.
30 De Campora E, Croce A, Bicciolo G, et al: Adenoid cystic carcinoma (cylindroma) of the pyriform sinus in paediatric age. Int J Pediatr Otorhinolaryngol 1987; 14:235–242.
31 Guzzo M, Ferrari A, Marcon I, et al: Salivary gland neoplasms in children: the experience of the Instituto Nazionale Tumori of Milan. Paediatr Blood Cancer 2006;47:806–810.
32 Kupfermann ME, de la Garza GO, Santillan AA, et al: Outcomes of pediatric patients with malignancies of the major salivary glands. Ann Surg Oncol 2010; 17:3301–3307.
33 Rahbar R, Grimmer JF, Vargas SO, et al: Mucoepidermoid carcinoma of the parotid gland in children: a 10 year experience. Arch Otolaryngol Head Neck Surg 2006;132:375–380.
34 Min R, Zun Z, Siyi L, et al: Gland-preserving surgery can effectively preserve gland function without increasing recurrence in treatment of benign submandibular gland tumour. Br J Oral Maxillofac Surg 2013;51:615–619.
35 Vedrine PO, Coffinet L, Temam S, et al: Mucoepidermoid carcinoma of salivary glands in the paediatric age group: 18 clinical cases, including 11 second malignant neoplasms. Head Neck 2006;28:827–833.
36 Thariat J, Vedrine P-O, Teman S, et al: The role of radiation therapy in paediatric mucoepidermoid carcinomas of the salivary glands. J Pediatr 2013;162:839–843.
37 Ribero K, Kowalski LP, Saba LM, et al: Epithelial salivary gland neoplasms in children and adolescents; a forty-four year experience. Med Pediatr Oncol 2002;39:594–600.

Patrick J. Bradley, MBA, FRCS
Emeritus Honorary Professor at the School of Medicine, The University of Nottingham
Honorary Consultant Head and Neck Oncological Surgeon, Nottingham University Hospitals
Queens Medical Centre Campus, Derby Road
Nottingham NG7 2UH (UK)

10 Chartwell Grove, Mapperley Plains
Nottingham NG3 5RD (UK)
E-Mail pjbradley@zoo.co.uk

Distant Metastases and Palliative Care

Tiffany A. Glazer[a] · Andrew G. Shuman[a,b]

[a]Department of Otolaryngology – Head and Neck Surgery and [b]Center for Bioethics and Social Sciences in Medicine, University of Michigan Medical School, Ann Arbor, Mich., USA

Abstract

Salivary gland neoplasms are rare and diverse tumors with variable disease courses, making it difficult to concisely summarize the management of distant metastases (DM). Nonetheless, there are trends of DM in salivary gland cancer that can be contextualized and reviewed. In general, the primary tumor characteristics that predict DM include the primary tumor site, tumor stage and grade, perineural spread, cervical nodal status, and genomic signatures. The most common site of DM is the lung, followed by the bone, liver, and brain. Depending on the clinical presentation, DM can be treated with watchful waiting, local therapies (surgery and/or radiation), or systemic therapy (cytotoxic or targeted chemotherapy). In general, DM confer a poor prognosis, and any cancer-directed treatment options should be carefully considered in the context of specific goals of care, symptom burden, and patient preference.

© 2016 S. Karger AG, Basel

Background

Scope of the Problem

Characterization of distant metastases (DM) in salivary gland cancer (SGC) is difficult owing to the rarity of the disease, histologic heterogeneity, and frequently prolonged clinical course. According to the World Health Organization (WHO), SGC accounts for just 0.3% of all cancers in the United States and <6% of all cancers of the head and neck. Additionally, the WHO classifies SGC into 24 distinct histologic varieties, many of which contain subtypes that are biologically distinct. Finally, DM in salivary cancer may manifest very different patterns of progression, ranging from rapidly growing disease leading to early demise to a much more gradual pattern of spread. Due to their slow growth, DM from salivary cancer in some patients may develop up to 3 decades after their primary diagnosis, and even among those who have developed DM, up to 10% survive for more than 10 years [1, 2]. These characteristics make it challenging to cohesively summarize the management of DM from salivary cancer.

Nonetheless, studies have reported that 11–40% of all patients with SGC will develop DM, with an overall 5-year distant recurrence-free probability (DRFP) of 73% and 10-year disease-free survival of 55% [2–5]. When patients were stratified according to risk group, the 5-year DRFP was around 90% for both low-risk tumors [acinic cell carcinoma (AcCC), low-grade mucoepidermoid carcinoma (MEC), cribriform adenocarcinoma of the tongue, and polymorphous low-grade adenocarcinoma] and intermediate-risk tumors (adenoid cystic carcinoma (ACC) and intermediate-grade MEC) and 54% for high-risk tumors [salivary duct carcinoma (SDC), high-grade MEC, carcinoma ex-pleomorphic adenoma, high-grade adenocarcinoma, and 'high-grade' carcinoma). When patients were stratified by clinical stage at initial diagnosis, the DFRPs for stage I, stage II, stage III, and stage IV tumors were 98%, 79%, 72%, and 38%, respectively [3] (table 1). Interestingly, multiple studies have shown that up to 64% of patients with DM had previously achieved locoregional control, which suggests that occult disease progression may have occurred at the time of initial diagnosis and treatment [5–8]. By taking into consideration a patient's specific clinical, histopathological, biochemical, and treatment variables, nomograms have been developed to predict individualized survival among patients with major SGC [9]. As the earlier chapters have described in detail, nomograms have benefits over the classic TNM staging system in that they incorporate multiple patient and tumor-related variables to help predict outcome for a specific patient.

Risk Factors
A review of the literature has suggested that DM tends to occur more commonly when primary tumors arise in the submandibular gland regardless of the histologic subtype, with an incidence of 37–41.9% [2, 8, 10, 11].

With regard to the clinical predictors of DM, it has been shown that the primary tumor stage, clinically positive cervical nodal (+cN) disease, overall clinical stage, and male gender are significant predictors of the 5-year DRFP in patients with SGC. In multivariate analysis, tumor stage, +cN disease, and male gender remained independently significant [3, 7]. One review cited that patients with cT4 disease were four times more likely ($p < 0.001$), those with +cN disease were 2.9 times more likely ($p = 0.004$), and males were 2.1 times more likely ($p = 0.018$) to develop DM [3]. Similarly, a review of parotid SGC found that the rates of DM in T1, T2, T3, and T4 tumors were 0%, 5%, 38%, and 73%, respectively ($p < 0.0001$) [7]. The postoperative nomogram created for patients with SGC concurs with these results and has shown that the cT stage, +cN disease, overall clinical stage, facial nerve paralysis, and skin involvement are predictive of survival outcome [9].

With regard to pathologic predictors, high-risk pathology, high-grade pathology, vascular invasion, perineural spread (PNS), positive surgical margins, the pT4 stage, positive pathologic nodal (+pN) disease, and the overall pathologic stage are significant predictors of the 5-year DRFP. In multivariate analysis, a high-grade pathology, PNS, and +pN disease were independent predictors of a worse outcome [3]. Another review focusing specifically on parotid cancer cited that the rates of DM were 2% for low-grade, 44% for intermediate-grade, and 36% for high-grade pathologies ($p <$

Table 1. Five-year DRFP for patients newly diagnosed with SGC stratified by primary tumor risk group and stage at the time of diagnosis

	5-year DRFP, %
Primary tumor risk group	
Low risk	89.8
Intermediate risk	88.5
High risk	53.7
Stage at initial diagnosis	
I	97.8
II	78.5
III	72.1
IV	38.1

Low-risk tumors include AcCC, low-grade MEC, myoepithelial carcinomas and polymorphous low-grade adenocarcinoma. Intermediate-risk tumors include ACC and intermediate-grade MEC. High-risk tumors include SDC, high-grade MEC, carcinoma ex-pleomorphic adenoma, high-grade adenocarcinoma, and 'high-grade carcinoma' [3].

0.001) and that patients with facial nerve involvement at diagnosis developed DM >50% of the time compared to 23% of those without facial nerve involvement. This study also showed that the incidence of DM in +pN disease was 68% versus 24% in –pN disease (p = 0.007) and suggested that multiple positive nodes were predictive of a higher risk of DM compared to a single positive cervical node (p = 0.014) [7]. The postoperative nomogram created for patients with SGC concurs with these results and has shown that the pT stage, +pN disease, PNS, vascular invasion, and the histologic grade are statistically significant predictors of survival outcome [9].

Focusing more specifically on high-risk and high-grade pathologies, the four subtypes of SGC that are the most likely to develop DM are all high-grade, with 25–64% of patients with undifferentiated 'high-grade carcinoma' developing DM, in addition to 52–82% of those with SDC, 42% of those with adenocarcinoma, 35% of those with squamous cell carcinoma, and 20–39% of those with carcinoma ex-pleomorphic adenoma [2, 3, 5]. By contrast, the risk of developing DM for intermediate- or low-risk tumors has been shown to be markedly lower, with an incidence of 16–17% for AcCC, 7–16% for low-grade MEC, and 6–17% for cribriform adenocarcinoma of the tongue [2, 3, 5].

Genetic mutations have been studied as potential predictors of DM, but the results are currently not well understood. Studies have reported that patients with ACC and high expression of either transmembrane protease, series 4 (TMPRSS40) (p < 0.001) or histone H3 lysine 9 trimethylation (H3K9me3) (p = 0.001) have an increased likelihood of DM [12, 13], while another study has shown that patients with SDC and p53 (p < 0.049) or HER-2/*neu* (p < 0.034) expression have an increased likelihood of DM [14]. While these genetic mutations may be helpful in estimating an individual patient's risk of developing DM, they cannot as yet be used to prognosticate individual patient outcomes or courses.

Prognosis with Distant Metastases
Because of the heterogeneity of metastatic SGC, it is difficult to accurately summarize patient survival outcomes. However, the majority of patients who die of SGC have DM at the time of death [2]. On average, the median overall survival time after the diagnosis of DM is 22–32 months (range, 4–90), but when bone, visceral, or vertebral metastases occur, the disease course is often more fulminant [4, 15]. The median survival time also varies widely depending on the patient and primary tumor characteristics [2]. For example, the overall survival after diagnosis of DM is lower for patients with a high-risk primary tumor compared to those with an intermediate- or low-risk primary tumor. In a large study of 565 patients with SGC, survival after diagnosis of DM in patients with high-risk pathology was 20 ± 4% at 1 year and 0% at 5 years [5]. In another series of 51 patients, all patients with high-grade, solid-type ACC died of DM [16]. In that same study, the overall survival rates after diagnosis of DM at 1 and 5 years for patients with AcCC were 80 ± 13% and 30 ± 14%, respectively, and for patients with ACC, the survival rates were 68 ± 7% and 32 ± 7%, respectively [5].

Site of Metastases
The most common site of DM in all SGCs is the lung, with incidences ranging from 49 to 91% for all DM [2, 3]. The next most common sites include the bone (13–40%), liver (4–19%), soft tissue (9%), distant nodal basins (8%), and brain (4–7%) (fig. 1). Up to 24% of patients who develop DM will manifest disease in more than one site, although this number may actually be higher since after DM are originally diagnosed, further metastatic workup and imaging may be deferred [2, 5, 8].

Diagnosis
The NCCN guidelines recommend standard head and neck cancer surveillance for SGC, which includes a regular clinical examination, post-treat-

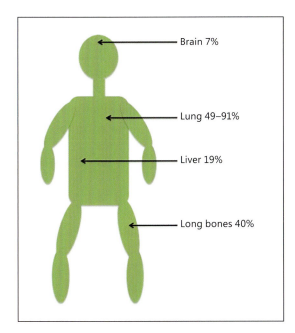

Fig. 1. Common sites of distant metastases in SGC. The most common site is the lung (49–91%), followed by the long bones (40%), liver (19%), soft tissue (9%), distant nodal basins (8%), brain (7%), kidney (2%), orbit (2%), pancreas (2%), and rarely, the vertebrae, skull, thyroid, ribs, ileum, retropharynx, and peritoneum [2, 3].

ment baseline imaging and further imaging (including chest imaging) as indicated based on the signs and symptoms [17]. Because the risk of developing DM remains for the duration of a patient's life after receiving a diagnosis of SGC, it is recommended that all patients obtain annual chest X-ray, which should be supplemented by chest CT when clinically indicated [2]. DM to the lung are typically the first and most common to develop, but patients rarely die of lung metastases, which makes the value of annual chest X-rays debatable [15]. Patients with DM who succumb to the disease frequently manifest metastases involving the long bones, vertebrae, or viscera [4, 15]. Supplemental tests, including biopsy, bone scan, liver function tests, and abdominal and brain scans, should be ordered when clinically indicated to detect DM at sites other than the lung [6].

Treatment Options

Defining Goals of Care

A diagnosis of DM typically heralds a transition away from treatment with curative intent. As such, the dynamics of the doctor-patient relationship and the rationale for cancer-directed therapy will necessarily evolve.

It is important for clinicians to provide honest insights into the patient's prognosis and expected trajectory to facilitate a productive, open-ended discussion with the patient and family to help define the goals of care [18]. The following components should be considered when discussing treatment options with a patient facing incurable cancer: acknowledge uncertainty; predict a general but realistic time frame; provide dynamic hope that can be reframed over time; recommend arranging affairs, including code status and advance directives; assure realistic availability during the patient's end of life transition; refer the patient for emotional and spiritual support; address the patient's goals; and encourage questions from the patient and family [19]. When each of those components is discussed in a frank and tactful manner, patients and families can better understand their prognosis and adjust the goals of care accordingly.

In general, treatment can be broadly categorized into cancer-directed therapies (surgery, radiation, and/or chemotherapy) and supportive care. It is important to note that these categories are by no means mutually exclusive. Even among patients receiving cancer-directed therapy, supportive management of bothersome symptoms is a priority. Likewise, even for patients prioritizing symptom management over treatment of their underlying cancer, cancer-directed therapies may well be necessary to confer the best chance for meaningful palliation. Regardless, the goals of treatment should be clarified when possible, whether they are aimed at the treatment of pain, prolongation of life, or some complex permutation thereof [4, 15].

Cancer-Directed Therapy of Distant Metastases
For selected patients with DM and good performance status, cancer-directed therapy via surgical resection, radiation, or chemotherapy is a reasonable consideration. For patients with symptomatic DM isolated to one site or when compromise of a vital structure is imminent, surgical resection or radiation may be considered, despite the lack of a proven survival benefit [20, 21]. In patients with solitary, indolent metastases, particularly those in the lung or those with few or no symptoms, the clinician must remember *primum non nocere* and acknowledge that watchful waiting may be the appropriate treatment option, particularly for slow-growing tumors [20]. Tantamount to these choices is consideration of the relative proportionality of the intervention based on its inherent burdens and potential benefits, weighed collectively by patient, family and clinician. In many cases, multidisciplinary input from a tumor board and other invested providers may be useful.

For patients who are unsuitable for surgery or chemotherapy due to either patient or oncologic factors and who have symptomatic disease or impending consequences of local tumor progression, palliative radiation therapy can be considered and may alleviate troublesome symptoms. The goals of palliative radiation therapy are to achieve tumor regression or stabilization and symptom control within a short overall treatment time with proportional side effects [22]. Hypofractionated external-beam radiation therapy in this population may offer reasonable compliance rates, good clinical efficacy, and acceptable toxicity and may be preferred over more invasive or extensive options by patients with incurable disease [23].

Palliative chemotherapy can be offered to patients with rapidly progressive or symptomatic DM when surgery or radiation is not possible and to those with multiple sites of disease. The most salient goal of systemic therapy is to prevent or delay the onset of disease-related symptoms, but it is unlikely to garner a defined survival benefit. If a patient does show a response, it is often disease stabilization rather than regression [1, 24]. Among the cytotoxic agents trialed in metastatic SGC, single agents often produce similar response rates as multi-agent regimens; in light of these similar response rates, one must consider the additional toxicity incurred to the patient with the use of multiple cytotoxic drugs [24].

HER2/*neu*, EGFR, and *c-kit* have shown overexpression in some SGCs, and their respective inhibitors have been trialed in patients with metastatic SGC. However, many of these trials have been small, making the results variable and difficult to interpret [24]. Because of the relative rarity, histologic heterogeneity, and prolonged clinical course of the disease, prospective clinical data on systemic therapy in SGC is limited and generally consists of case reports and small phase II trials of patients with a variety of histologic subtypes, all of which inherently behave differently.

The NCCN guidelines recommend enrolling eligible, motivated patients with metastatic SGC into a clinical trial when appropriate, as dictated by patient preference, disease status, medical co-morbidities and performance status [17]. However, it is important not to 'oversell' the trial or experimental agent and provide false hope, thereby creating a therapeutic misconception [18]. Clinicians should engage in an honest, open-ended, patient-centric rather than cancer-centric conversation when considering enrolling a patient in a clinical trial. This includes reviewing the objectives of the trial and its plausible advantages and disadvantages, including potential adverse effects, to help the patient to make a preference-sensitive decision [25].

Supportive Care
Many patients with metastatic SGC are not appropriate candidates for cancer-directed therapies. Patients of advanced age who are frail with severe comorbidities, a poor performance status, and disseminated, rapidly progressive or large-

volume metastatic disease may have a predicted survival time of less than 1 year regardless of the treatment employed [10]. Such patients may not benefit from cancer-directed therapy as the burdens become disproportionate to the intended benefit [18]. As previously stated, however, symptom control and cancer treatment need not be mutually exclusive.

Once the decision has been made by the patient to prioritize symptom management rather than cancer-directed treatment, best supportive care should focus on the patient's specific goals, as well as managing pain, treating bothersome symptoms, and providing the means to live with dignity and respect, regardless of the anticipated prognosis. While each patient will have a unique symptom burden and will therefore require individually tailored palliation, nearly every terminal cancer patient will experience pain, making experts in pain control invaluable in personalizing an optimal regimen for these patients. Some of the strongest predictors of adequate palliation are proper pain control and the patient's perception that their pain is being respected and addressed by caregivers [25].

Early involvement of palliative care consultants in the treatment of these patients is critical, as optimal pain and symptom management should be established well before the terminal phase of a patient's disease. This may or may not involve hospice services, either in the home or in a facility. In addition, insurance and cultural variables, including access to care and availability of services, will undoubtedly influence how end-of-life care is delivered in specific settings. While many patients might think of palliative care as 'giving up,' patients and families succumbing to head and neck cancer who are involved with palliative care experts have significantly better symptom control and care at the time of death [25]. For patients with terminal lung cancer, not only has early palliative care consultation been shown to result in a better quality of life, but the patients tend to have fewer depressive symptoms, less aggressive care at the end of life, and intriguingly, longer survival [26]. The importance of early and appropriate palliative care cannot be overstated.

Conclusion

In summary, SGC comprises a small proportion of head and neck cancers worldwide and has a variable disease course. Prognosis is highly variable and depends on the specific patient and tumor factors. Treatment should be based on both the disease characteristics and the patient-directed goals of care. Select patients with DM should undergo cancer-directed therapy, which may involve surgical resection, radiation, and/or chemotherapy. For patients who are not candidates for and/or who decline these options, best supportive care should commence with early palliative care involvement and an emphasis on pain and symptom control. Honest and compassionate conversations among clinicians and patients are obligatory to identify realistic goals and to make individualized end-of-life decisions for patients with incurable cancer and frequently variable disease trajectories.

References

1. Laurie SA, Licitra L: Systemic therapy in the palliative management of advanced salivary gland cancers. J Clin Oncol 2006;24:2673–2678.
2. Bradley PJ: Distant metastases from salivary glands cancer. ORL J Otorhinolaryngol Relat Spec 2001;63:233–242.
3. Ali S, Bryant R, Palmer FL, et al: Distant metastases in patients with carcinoma of the major salivary glands. Ann Surg Oncol 2015, in press.
4. Teo PM, Chan AT, Lee WY, et al: Failure patterns and factors affecting prognosis of salivary gland carcinoma: retrospective study. Hong Kong Med J 2000;6:29–36.

5 Terhaard CH, Lubsen H, Van der Tweel I, et al: Salivary gland carcinoma: independent prognostic factors for locoregional control, distant metastases, and overall survival: results of the Dutch head and neck oncology cooperative group. Head Neck 2004;26:681–692; discussion 692–693.

6 Gallo O, Franchi A, Bottai GV, et al: Risk factors for distant metastases from carcinoma of the parotid gland. Cancer 1997;80:844–851.

7 Renehan AG, Gleave EN, Slevin NJ, et al: Clinico-pathological and treatment-related factors influencing survival in parotid cancer. Br J Cancer 1999;80:1296–1300.

8 Mariano FV, da Silva SD, Chulan TC, et al: Clinicopathological factors are predictors of distant metastasis from major salivary gland carcinomas. Int J Oral Maxillofac Surg 2011;40:504–509.

9 Ali S, Palmer FL, Yu C, et al: Postoperative nomograms predictive of survival after surgical management of malignant tumors of the major salivary glands. Ann Surg Oncol 2014;21:637–642.

10 Schwentner I, Obrist P, Thumfart W, et al: Distant metastasis of parotid gland tumors. Acta Otolaryngol 2006;126:340–345.

11 Speight PM, Barrett AW: Prognostic factors in malignant tumours of the salivary glands. Br J Oral Maxillofac Surg 2009;47:587–593.

12 Xia R, Zhou R, Tian Z, et al: High expression of H3K9me3 is a strong predictor of poor survival in patients with salivary adenoid cystic carcinoma. Arch Pathol Lab Med 2013;137:1761–1769.

13 Dai W, Zhou Q, Xu Z, et al: Expression of TMPRSS4 in patients with salivary adenoid cystic carcinoma: correlation with clinicopathological features and prognosis. Med Oncol 2013;30:749.

14 Jaehne M, Roeser K, Jaekel T, et al: Clinical and immunohistologic typing of salivary duct carcinoma: a report of 50 cases. Cancer 2005;103:2526–2533.

15 van der Wal JE, Becking AG, Snow GB, et al: Distant metastases of adenoid cystic carcinoma of the salivary glands and the value of diagnostic examinations during follow-up. Head Neck 2002;24:779–783.

16 Koka VN, Tiwari RM, van der Waal I, et al: Adenoid cystic carcinoma of the salivary glands: clinicopathological survey of 51 patients. J Laryngol Otol 1989;103:675–679.

17 Pfister DG, Ang KK, Brizel DM, et al: Head and neck cancers, version 2.2013. Featured updates to the NCCN guidelines. J Natl Compr Canc Netw 2013;11:917–923.

18 Shuman AG, Fins JJ, Prince ME: Improving end-of-life care for head and neck cancer patients. Expert Rev Anticancer Ther 2012;12:335–343.

19 Loprinzi CL, Johnson ME, Steer G: Doc, how much time do I have? J Clin Oncol 2000;18:699–701.

20 Fordice J, Kershaw C, El-Naggar A, et al: Adenoid cystic carcinoma of the head and neck: predictors of morbidity and mortality. Arch Otolaryngol Head Neck Surg 1999;125:149–152.

21 Bobbio A, Copelli C, Ampollini L, et al: Lung metastasis resection of adenoid cystic carcinoma of salivary glands. Eur J Cardiothoras Surg 2008;33:790–793.

22 Garden AS, Weber RS, Ang KK, et al: Postoperative radiation therapy for malignant tumors of minor salivary glands. Outcome and patterns of failure. Cancer 1994; 73: 2563–2569.

23 Al-mamgani A, Tans L, Van rooij PH, et al: Hypofractionated radiotherapy denoted as the 'Christie scheme': an effective means of palliating patients with head and neck cancers not suitable for curative treatment. Acta Oncol 2009;48:562–570.

24 Lagha A, Chraiet N, Ayadi M, et al: Systemic therapy in the management of metastatic or advanced salivary gland cancers. Head Neck Oncol 2012;4:19.

25 Shuman AG, Yang Y, Taylor JM, et al: End-of-life care among head and neck cancer patients. Otolaryngol Head Neck Surg 2011;144:733–739.

26 Temel JS, Greer JA, Muzikansky A, et al: Early palliative care for patients with metastatic non-small-cell lung cancer. N Engl J Med 2010;363:733–742.

Andrew G. Shuman, MD
1500 E. Medical Center Drive
Taubman 1904
Ann Arbor, MI 48109 (USA)
E-Mail shumana@med.umich.edu

Quality of Life after Salivary Gland Surgery

Mark K. Wax[a] · Yoav P. Talmi[b]

[a]Department of Otolaryngology-HNS, Oregon Health and Sciences University, Portland, Oreg., USA; [b]Head and Neck Service, Department of Otolaryngology, Head and Neck Surgery, The Chaim Sheba Medical Center, Tel Hashomer, Israel

Abstract

Quality of life (QoL) has been recognized as an important endpoint in addition to disease-related and global survival. It is particularly important for patients with salivary gland neoplastic disease. For patients who are undergoing benign salivary gland tumor surgery, cosmetic and functional outcomes are extremely important, as these patients' psychological well-being and ability to function in society can be severely impacted. The following issues related to surgical treatment are discussed: incision, loss of local tissue sensation, development of Frey's syndrome, facial nerve function, and cosmesis. Improvements in the placement of the incision combined with additional minimally invasive procedures have improved QoL. The ultimate goal of benign parotid neoplastic surgery is complete tumor excision while avoiding cosmetic and functional damage, which includes preservation of the function of the facial nerve and its branches; this is the key to maintaining preoperative levels of QoL. There are many measures available to improve cosmesis that have minimal morbidity and that, when used, can provide significant improvements in patient outcomes. The treatment of malignant salivary gland neoplasms is primarily directed at treating the malignancy. When surgical treatment affects important neighboring structures, such as the lingual or hypoglossal nerves, as in submandibular/sublingual cancer, there is a tremendous effect on QoL if postoperative dysfunction of these structures results. Often, this treatment involves using ancillary surgical procedures, such as neck dissection, or nonsurgical treatment, such as radiation therapy. The effect of such multi-modality treatment on QoL is significant. The treatment of underlying salivary disease is often overshadowed by these adjunctive treatments.

© 2016 S. Karger AG, Basel

Introduction

Health-related quality of life (QoL) refers to a patient's perception of the effects of a disease and treatment and their impact on daily functioning and general feelings of well-being. The World Health Organization defines QoL as an 'individual's perception of his or her position in life in the context of the culture and value systems in which the patient lives and in relation to his or her goals, expectations, standards, and concerns'. QoL can be defined as the 'perceived discrepancy between the reality of what a person has and the concept of what the person wants, needs, or expects' [1]. QoL has two fundamental foundations: (1) it is multidimensional, incorporating physical, psychological, social, emotional, and functional domains, and (2) it is subjective and must be based on self-reporting according to a patient's own experiences. QoL is recognized as an important endpoint in clinical research, in addition to time-honored endpoints, such as disease-related and global survival. QoL is particularly relevant for patients with tumors of the head and neck because social interaction and expression depend largely on the structural and functional integrity of the head and neck area. Thus, data from QoL questionnaires are becoming an important supplement to information pertaining to treatment outcome. There are now a variety of well-validated QoL instruments available for use in the fields of head and neck surgery as well as oncology. These include generic instruments and tumor site-specific instruments.

Naturally, as a majority of salivary gland procedures are used for benign disease, a shift in considerations regarding QoL may ensue. Factors that may seem insignificant to a person with malignant disease, even after such disease is deemed cured, may be important to a basically healthy individual. It should be remembered that the issues faced by an individual with benign disease are also faced by the patient with malignant disease. It is often the case that a treatment paradigm is more extensive for a patient with malignant disease. However, this was not found to be the case in a QoL study on parotidectomy [2]. Indeed, Beutner et al. [3] reported no alterations in QoL 1 year after lateral or superficial parotidectomy. Fang et al. [4] evaluated long-term QoL in children who had survived parotid tumor surgery and reported that such treatment had a limited negative impact.

While tumors of the sublingual salivary glands are rare, tumors of the submandibular glands are not uncommon. Reported QoL issues associated with the submandibular glands mainly involve drooling and relocation [5]; however, QoL issues pertaining to submandibular gland tumor surgery have not been addressed in the literature. If surgery is complicated by local nerve damage, then there is a tremendous effect on QoL. Uncomplicated submandibular or sublingual gland surgery is not reported to have an effect on QoL.

The effects of the treatment of malignant minor salivary glands or submandibular glands are hard to interpret. The literature is scant, and the majority of cases reported have received some form of ancillary treatment. Thus, the effect of local resection is subsumed by the effect of ancillary treatment, radiation therapy, and extensive surgical resection. Therefore, unfortunately, there is not much to be said about the effects of treatment on malignant tumors other than the parotid gland.

This chapter will discuss QoL as it relates to the surgical treatment of salivary gland pathology. Issues directly related to such surgical treatment include the cutting of an incision, the perception of sensation, the development of Frey's syndrome, the status of facial nerve function and the maintenance of cosmesis.

Sensation

Patel et al. [6] have stated that while many patients experience sensory deficits, overall QoL is not significantly affected after great auricular nerve (GAN) sacrifice during parotidectomy. Patients who report multiple abnormal sensations, however, would benefit from additional counseling and from reassurance that the number of sensations will diminish with time [6]. Yokoshima et al. [7] attempted preservation of the posterior branch of the GAN in parotidectomy, succeeding in 26 of 40 patients (65%). Scores for QoL were significantly higher in the group of patients with GAN preservation at 2, 3 and 6 months. Ryan and Fee [8] described 27 patients who underwent parotidectomy with GAN sacrifice. The prevalence and average area of anesthesia decreased continually during the first year according to sensory testing and patient scoring. Half of the patients had no anesthesia at 12 months. While the impact of GAN sacrifice morbidity on patient QoL was reported as tolerable and improved during the first postoperative year, the patients still had difficulties using the telephone, shaving, combing their hair, wearing earrings, and sleeping on the operated side because of both anesthesia and paresthesia. Conversely, Min et al. [9] evaluated the necessity of preserving the posterior branch of the GAN in 24 of 46 prospectively analyzed patients using a QoL questionnaire. Sensory deficits improved over time in both groups, and after 12 months only minimal sensory loss remained. QoL was not significantly different between the groups, and it was concluded that preservation of the posterior branch of the GAN might not be necessary. George et al. [10] assessed evidence related to preservation of the GAN in parotidectomy with regard to morbidity and QoL in a systematic review of prospective and retrospective studies in the English literature. The outcome measures were tactile sensation, pain, thermal sensitivity, and QoL. Although QoL does not seem to be adversely affected when the GAN is sacrificed, preservation of the posterior branch was recommended in 8 studies. Level Ib evidence shows that preservation of the GAN minimizes postoperative sensory disturbances and should be considered whenever tumor clearance is not compromised. No evidence was found to support that overall QoL is affected when the GAN is sacrificed in a report by Grammatica et al. [11]. Contrary to some reports, the presence of the posterior branch of the GAN is not readily observed, and in some cases its presence is questionable. Yet, good surgical practice dictates the preservation of any structure for which removal is not warranted.

Frey's Syndrome

de Bree et al. [12] discussed the etiology, incidence, treatment and prevention of Frey's syndrome occurring after parotid surgery and commented that only a minority of patients need treatment for this problem. Hartl et al. [13] measured patient-reported QoL before and after botulinum toxin A treatment of Frey's syndrome. Patient-reported functional QoL improved significantly, although their social and emotional scores were not significantly modified. The benefit lasted for over 1.5 years in 60% of the patients. Baek et al. [14] studied long-term QoL issues after parotidectomy. Frey's syndrome was identified as the most serious self-perceived sequela, and the resulting discomfort worsened with time. Subjective perception of Frey's syndrome was most serious in total parotidectomy cases, from 12 months to over 5 years following surgery. The authors concluded that additional measures that prevent or ameliorate Frey's syndrome are likely to improve long-term QoL after parotidectomy.

Ciuman et al. [15] documented the outcomes and impact on general and symptom-specific QoL after various types of parotid resection. General QoL was assessed based on both global health status and global QoL scales from the European Organization for Research and Treatment of Cancer QoL Questionnaire. Symptom-specific QoL was assessed using the Parotidectomy Outcome Inventory-8, and aesthetic outcome was evaluated using an ordinal scale. The outcome of parotidectomy in benign disease has little impact on general QoL and global health status. However, hypoesthesia or dysesthesia, Frey's syndrome, and cosmetic discontent were found to be quite common and may affect symptom-specific and general QoL. Gunsoy et al. [16] evaluated QoL after parotidectomy in 49 patients. Patients with symptomatic Frey's syndrome had statistically significantly decreased scores related to social functions, economic difficulties, speech defects, reduced sexuality, and nutritional parameters. In an early study of QoL and parotidectomy, 30 patients (57%) reported local erythema and/or sweating during eating. Although the overall score was rather high, the

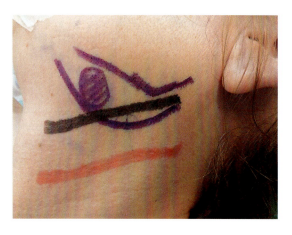 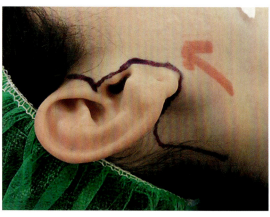

Fig. 1. Open transcervical approach. A patient with a 1.2 cm pleomorphic adenoma of the submandibular gland. The lower purple line indicates a 'classic' incision, while the black and red lines follow skin creases.

Fig. 2. Planned parotidectomy incision illustrating a modified facelift approach.

importance attributed to this problem was low. Two patients defined these symptoms as intolerable, but no patient received treatment or was interested in seeking treatment with botulinum toxin A.

Incisions

Submandibular Gland

The classic incision is described as two fingerbreadths below the mandible, and this is seemingly the approach of many surgeons [17]. This placement of an incision is suggested in order to prevent damage to the mandibular branch of the facial nerve. Described methods of excision include open, endoscopic, and robotic-assisted resections. The corresponding surgical approaches vary and may incorporate transcervical, submental, transoral or retroauricular techniques in order to reduce or eliminate discernible scarring [18]. A logical approach allowing for both adequate access and cosmesis would be to place an incision in a suitable skin fold, even if not adhering to the '2-finger' rule (fig. 1).

Parotid Gland

The classic incision was popularized by Blair in 1936, and since then a variety of incisions has been offered [19]. Terris et al. [20] suggested a modified face-lift incision that resulted in improved patient satisfaction compared with the modified Blair incision (fig. 2, 3). Jost et al. [21] suggested what they defined as a plastic approach to parotidectomy. This included using a four-step procedure including a face-lift incision. Sternocleidomastoid muscle transposition combined with face-lift incision was suggested by Chow et al. [22] to improve the cosmetic outcome of superficial parotidectomy, with satisfactory results. In one study, surprisingly, scores for appearance were significantly better for the younger age group compared with those over 45 years of age [2]. As the incisions were similar, it could have been expected that the younger patients would be more concerned with changes in their appearance. Perhaps reduced facial tone in the older age group serves to amplify the effect of scarring.

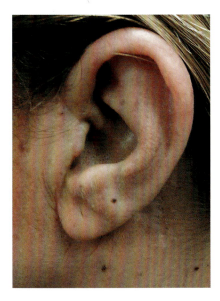

Fig. 3. A view of a modified facelift incision 1 year following parotid surgery.

Fig. 4. The use of a sternocleidomastoid flap for augmentation of the tail of a parotid resection and the use of an acellular matrix to augment the superior parotid defect.

Cosmesis

The parotid gland functions as soft tissue filler in the lateral aspect of the cheek and the lateral mandibular area. Removal of the parotid gland can lead to a minor or significant concave deformity. The extent of the deformity will be dependent on the pathological process that warrants parotid excision. Historically, this area has not been reconstructed. The majority of patients are left with a concave defect that tends to flatten out over time, although it is always present. Prevention of injury to the facial nerve is probably the primary concern in parotid surgery; consequently, cosmetic deformity after parotid resection is considered by most to be a relatively minor issue in the immediate postoperative period. As time progresses, patients learn to live with and accept the defect. Ciuman et al. [15] demonstrated that surgery for benign parotid disease had little impact on general QoL or global health status, but that cosmetic discontent exists quite frequently in this patient population. They also demonstrated that patients were significantly more satisfied after partial compared to superficial or total parotidectomy. The lack of symptom-specific questions on the QoL questionnaire limited the ability to evaluate this issue. Bianchi et al. [23] reviewed 274 patients and found that the most essential factor that impacted aesthetic outcome was the amount of parotid tissue removed.

The reconstruction paradigm for soft tissue defects of the parotid ranges from sternocleidomastoid flaps to free tissue transfer [24]. All of these reconstructive interventions can provide varying volumes of tissue to fill in a defect and can be tailored to match both its volume and consistency. Small tissue defects have been shown to be well reconstructed with either allogeneic acellular dermis or local soft tissue. The superficial musculoaponeurotic system (SMAS) and the sternocleidomastoid muscle flaps have all been used to reconstruct parotidectomy defects. The use of a sternocleidomastoid flap has helped improve aesthetic outcomes [22]. Bianchi et al. [23] demonstrated that the use of a soft-tissue flap, in their case, the sternocleidomastoid or SMAS, resulted in a significant improvement in patient satisfaction with their final appearance. Both the

SMAS and the sternocleidomastoid flap were equivalent as soft-tissue filler in this study. Both of these reconstructive modalities are good for small-volume soft-tissue defects. When larger defects result from parotid gland surgery, one must turn to other soft-tissue volume fillers (fig. 4). While it seems intuitive that larger volume defects result in larger concavity and thus a more significant facial deformity, the effects on QoL from either not reconstructing or reconstructing these larger defects have not been reported. Results and improvements in patient satisfaction are mainly anecdotal, with little evidence above level IV.

It is intuitive that a patient's QoL is better following a small parotid lesion excision involving minimal salivary tissue excision that requires minimal or no reconstruction than that of a patient with a large parotid lesion producing a large defect that requires a large volume of soft tissue. The former scenario is likely to result in a better-than-anticipated cosmetic result and therefore a better QoL score than the latter scenario. Cannady et al. [25] described such a scenario in a series of 18 patients, all of whom underwent free tissue transfer for soft-tissue volume reconstruction of a contour defect following total parotidectomy. All patients reported an acceptable cosmetic outcome.

Facial Paralysis

Earlier in this chapter, we discussed that the larger the extent of a parotid resection, the greater effect it has on patient QoL. Adding any technical factors that result in facial paralysis will have a significant effect on patient QoL that will override all of the other discussed issues. Facial paralysis is a devastating complication or sequela of parotidectomy. The facial nerve is responsible not only for a patient's perception of themselves but also for many physiologic processes, such as the elevation of the brow to maximize upward gaze, the ability to open and close the eyes to maintain corneal lubrication, and the maintenance of adequate muscular tension of the lower lid to allow free flow and drainage of tear fluid. All are important roles of a normally functioning facial nerve in the upper face. In the mid face, the facial nerve is responsible for adequate nasal valve expansion to maintain the nasal passage. Finally, in the lower third of the face, movement of the lateral commissure and maintenance of oral competence are of paramount importance for talking, eating and drinking. Whenever facial nerve function is interfered with, whether transiently or permanently, the effect from the physiological as well as psychological perspective will be devastating depending on which branches are affected. Reconstruction of the facial nerve is of paramount importance and should be done either at the time of resection or as soon after it is noted.

When the facial nerve is repaired, excellent return of function can be expected, with many authors reporting a House-Brackman Grade –3 result [26]. While the functional results are excellent and the patient is protected from the devastating effects of ocular or oral dysfunction, QoL parameters still suffer. Patients will have significant disturbances in social eating, problems opening the mouth, and issues with sexuality [26]. Social function and global mental health are also lower than in the general population. Henstrom et al. [27] reviewed 49 patients presenting with paralytic lagophthalmos. These patients were treated with static periocular reanimation. Overall QoL after the intervention, as measured by the Faculty and Course Evaluation instrument, demonstrated an increase from a baseline of 44–52 ($p < 0.001$). Symptomatically, patients were much improved, with decreases in dryness, irritated and scratchy feelings in the eye ($p < 0.001$). Soler et al. [28] reviewed 28 patients who underwent facial nerve resection and evaluated them using the Nasal Obstruction Septoplasty Evaluation tool. Eighteen of these patients underwent immediate nasal valve reconstruction, while 10 did not have nasal valve reconstruction. The patients who did not undergo reconstruction demonstrated significantly more congestion or stuffiness, nasal blockage or congestion, trouble breathing through the nose, and trouble sleeping, as well as inability to intake enough air during exertion ($p < 0.05$ for all). At a median follow-up of 2 years, cosmesis was acceptable, as were the functional outcomes in the group that underwent the intervention. Lindsay et al. [29] reviewed 20 patients with nonflaccid facial paralysis who underwent free tissue transfer reconstruction using the gracilis muscle. A significant increase in the Faculty and Course Evaluation score was seen after free muscle transfer. Using a facial nerve grading scale, significant improvements in smile score, lower lip length at rest and smile symmetry were observed in the group that underwent the intervention. There was a quantifiable improvement in QoL in this group.

One issue that requires highlighting is the timing of surgical rehabilitative intervention. Should a reconstructive procedure be performed during surgical ablation or after a patient has recovered? It is clear that the longer a patient has a sequela of facial paralysis the more their QoL is affected. Watts et al. [30] reported 52 patients who underwent immediate rehabilitation of the orbital complex at the time of facial nerve resection. The eye complex was rehabilitated with placement of a gold weight in the upper lid of all 52 patients, and a lateral tarsal strip procedure was used in 48 of the 52 patients. Greater than 80% of the patients had lasting effect and required no revision surgery for the orbital complex. They concluded that immediate reconstruction and rehabilitation of the orbital complex at the time of facial nerve resection should be undertaken.

Therefore, if surgical intervention is considered necessary to improve cosmesis and/or physiology, then such interventions should target subunit of the face, regardless of whether it is dynamic, and can be tailored to improve patient QoL.

Salivary Gland Carcinoma

Malignant neoplasms of the minor salivary glands are relatively uncommon. When they occur, treatment involves surgical resection of the gland with varying amounts of surrounding tissue. QoL parameters in this instance will have to take into consideration the amount and size of the underlying composite tissue resection. For example, patients with minor salivary gland neoplasms of the lip will have issues with lip resection and reconstruction, and patients with minor salivary gland tumors of the hard palate will have major issues with palatal resection. The effects on QoL in these cases will be most dramatically related to the quantity of oral tissues resected and not the loss of the minor salivary gland per se.

Malignant lesions of the major salivary glands other than the parotid gland suffer from the same issues as those related to the minor salivary glands. Surgical resection frequently involves neighboring or adjacent structures and often involves the use of ancillary treatment, such as radiation therapy. Thus, the effects on QoL are associated with the local effects caused by irradiation to the head and neck and/or with the amount of local tissue resected. A search of the literature finds that the rarity of these tumors combined with the adjacent structures resected leads to a paucity of studies on the direct effects of resection of the gland. What can be discerned is that the effect of the treatment on QoL is the same as that for the treatment of head and neck cancers in the same area.

Parotid cancer is different in that is has a high enough incidence to have been studied specifically. In localized forms, management will be the same as that used for larger benign disease. These patients will face the same issues as those who undergo resection for benign disease; these issues are related to the cutting of the incision, the perception of sensation, the development of Frey's syndrome, the status of facial nerve function and the maintenance of cosmesis. Studies have demonstrated that the effects on QoL are the same between this population (as related to the above-described areas) and patients with benign disease [3, 4].

Treatment of patients with parotid cancer most often involves surgical resection of the parotid gland, often with some form of neck dissection and adjuvant therapy as needed. The defect may be larger than that for benign disease, and the facial nerve is at more risk of resection or damage. As expected, these factors play a role in the effect of QoL for benign disease. When they are exacerbated in the treatment of malignant disease, the effect on QoL is magnified [31].

Finally, the effect of postoperative radiation therapy for parotid cancer has been reported by Al-Mamgani et al. [32]. They reviewed 186 treated patients, 47 of whom had a minimum follow-up of more than 12 months, with complete data endpoints available. This study demonstrated that there was a significant decline in QoL over the course of treatment, but most patients returned to a pretreatment level around 6 weeks. Specific measurements of taste, sticky saliva, and dry mouth took longer to return to baseline. Baseline patient symptoms in this study were recorded at the commencement of radiation therapy thus, there are no data reported on the effect of multimodality treatment from the pretreatment phase.

Conclusion

There has been an expansion in the use of QoL measures in patients undergoing parotid surgery. The paradigm for reconstruction as well as the extent of resection continues to evolve as these measures of outcomes demonstrate improvements not only in oncologic and surgical outcomes but also in rehabilitation and QoL. As in many areas of head and neck surgery, functional and rehabilitative outcomes are now a dominant factor when considering the extent of a surgical resection and the type and timing of reconstructive modalities employed.

References

1 Morton RP, Izzard ME: Quality-of-life outcomes in head and neck cancer patients. World J Surg 2003;27:884–889.
2 Nitzan D, Kronenberg J, Horowitz Z, et al: Quality of life following parotidectomy for malignant and benign disease. Plast Reconstr Surg 2004;114:1060–1067.
3 Beutner D, Wittekindt C, Dinh S, et al: Impact of lateral parotidectomy for benign tumors on quality of life. Acta Otolaryngol 2006;126:1091–1095.
4 Fang QG, Shi S, Zhang X, Li M, et al: Long term quality of life in pediatric patients surviving parotid tumors. Int J Pediatr Otorhinolaryngol 2014;78:235–237.
5 Syeda F, Ahsan F, Nunez DA: Quality of life outcome analysis in patients undergoing submandibular duct repositioning surgery for sialorrhoea. J Laryngol Otol 2007;121:555–558.
6 Patel N, Har-El G, Rosenfeld R: Quality of life after great auricular nerve sacrifice during parotidectomy. Arch Otolaryngol Head Neck Surg 2001;127:884–888.
7 Yokoshima K, Nakamizo M, Ozu C, et al: Significance of preserving the posterior branch of the great auricular nerve in parotidectomy. J Nippon Med Sch 2004;71:323–327.
8 Ryan WR, Fee WE Jr: Great auricular nerve morbidity after nerve sacrifice during parotidectomy. Arch Otolaryngol Head Neck Surg 2006;132:642–649.
9 Min HJ, Lee HS, Lee YS, et al: Is it necessary to preserve the posterior branch of the great auricular nerve in parotidectomy? Otolaryngol Head Neck Surg 2007;137:636–641
10 George M, Karkos PD, Dwivedi RC, et al: Preservation of greater auricular nerve during parotidectomy: sensation, quality of life, and morbidity issues. A systematic review. Head Neck 2014;36:603–608.
11 Grammatica A, Perotti P, Mancini F, et al: Great auricular nerve preservation in parotid gland surgery: long-term outcomes. Laryngoscope 2015;125:1107–1112.
12 de Bree R, van der Waal I, Leemans CR: Management of Frey syndrome. Head Neck 2007;29:773–778.
13 Hartl DM, Julieron M, Le Ridant AM, et al: Botulinum toxin A for quality of life improvement in post-parotidectomy gustatory sweating (Frey's syndrome). J Laryngol Otol 2008;122:1100–1104.
14 Baek CW, Chung MK, Jeong HS, et al: Questionnaire evaluation of sequelae over 5 years after parotidectomy for benign diseases. J Plast Reconstr Aesthet Surg 2009;62:633–638.

15 Ciuman RR, Oels W, Jaussi R, et al: Outcome, general, and symptom-specific quality of life after various types of parotid resection. Laryngoscope 2012;122:1254–1261
16 Gunsoy B, Vuralkan E, Sonbay ND, et al: Quality of life following surgical treatment of benign parotid disease. Indian J Otolaryngol Head Neck Surg 2013;65(suppl 1):105–111.
17 Smyth C, Jackson C, Smith C, et al: Should we use fingerbreadth measurements in submandibular gland surgery? A critical appraisal of the technique. J Laryngol Otol 2014;128:932–934.
18 Beahm DD, Peleaz L, Nuss DW, et al: Surgical approaches to the submandibular gland: a review of literature. Int J Surg 2009;7:503–509.
19 Ferreria JL, Mauriño N, Michael E, et al: Surgery of the parotid region: a new approach. J Oral Maxillofac Surg 1990;48:803–807.
20 Terris DJ, Tuffo KM, Fee WE Jr: Modified facelift incision for parotidectomy. J Laryngol Otol 1994;108:574–578.
21 Jost G, Guenon P, Gentil S: Parotidectomy: a plastic approach. Aesthetic Plast Surg 1999;23:1–4.
22 Chow TL, Lam CY, Chiu PW, et al: Sternocleidomastoid-muscle transposition improves the cosmetic outcome of superficial parotidectomy. Br J Plast Surg 2001;54:409–411.
23 Bianchi B, Ferri A, Ferrari S, et al: Improving esthetic results in benign parotid surgery: statistical evaluation of face-lift approach, sternocleidomastoid flap, and superficial musculoaponeurotic system flap application. J Oral Maxillofac Surg 2011;69:1235–1241.
24 Militsakh ON, Sanderson JA, Lin D, et al: Rehabilitation of a parotidectomy patient–a systematic approach. Head Neck 2013;35:1349–1361.
25 Cannady SB, Seth R, Fritz MA, et al: Total parotidectomy defect reconstruction using the buried free flap. Otolaryngol Head Neck Surg 2010;143:637–643.
26 Guntinas-Lichius O, Straesser A, Streppel M: Quality of life after facial nerve repair. Laryngoscope 2007;117:421–426.
27 Henstrom DK, Lindsay RW, Cheney ML, et al: Surgical treatment of the periocular complex and improvement of quality of life in patients with facial paralysis. Arch Facial Plast Surg 2011;13:125–128.
28 Soler ZM, Rosenthal E, Wax MK: Immediate nasal valve reconstruction after facial nerve resection. Arch Facial Plast Surg 2008;10:312–315.
29 Lindsay RW, Bhama P, Hadlock TA: Quality-of-life improvement after free gracilis muscle transfer for smile restoration in patients with facial paralysis. JAMA Facial Plast Surg 2014;16:419–424.
30 Watts TL, Chard R, Weber SM, et al: Immediate eye rehabilitation at the time of facial nerve sacrifice. Otolaryngol Head Neck Surg 2011;144:353–356.
31 Stenner M, Beenen F, Hahn M, et al: Exploratory study of long-term health-related quality of life in patients with surgically treated primary parotid gland cancer. Head Neck 2016;38:111–117.
32 Al-Mamgani A, van Rooij P, Verduijn GM, et al: Long-term outcomes and quality of life of 186 patients with primary parotid carcinoma treated with surgery and radiotherapy at the Daniel den Hoed Cancer Center. Int J Radiat Oncol Biol Phys 2012;84:189–195.

Mark K. Wax, MD, Professor of Otolaryngology and Maxillofacial Surgery
Department of Otolaryngology-HNS, Oregon Health and Sciences University
3181 SW Sam Jackson Park Road PV-01
Portland, OR 97239 (USA)
E-Mail waxm@ohsu.edu

Salivary Gland Neoplasms: Future Perspectives

As is evident from the excellent chapters in this volume, many advances in the diagnosis and management of salivary gland neoplasms have occurred in the past several decades that have resulted in improved surgical outcomes and less morbidity. These advances have translated into improved quality of life. Further advances in the near future are certain. This chapter will present our perspectives on potential future advances that will further impact the care of the patient with a salivary gland neoplasm.

A great deal of new information has been learned about the pathogenesis of salivary gland neoplasms. Although radiation is the primary risk factor for the development of salivary gland neoplasms, the Epstein-Barr virus also plays a role. Other viruses may also be identified. The molecular events that lead to salivary gland tumorigenesis are becoming clearer. Due the diverse nature of salivary gland neoplasms and further diversity among tumor types, more molecular investigations are needed to further understand the complex molecular biology of these tumors, including the cytogenetic and epigenetic alterations and the heterogeneous pathway processes that occur, leading to salivary gland tumorigenesis. These basic science advances will hopefully translate into preventative and early detection strategies, refined evaluation techniques, and novel therapeutic approaches.

The diagnosis of salivary gland neoplasms will be facilitated by advances in imaging methods. Recently, there has been steady adoption of in-office ultrasound among head and neck surgeons. This technology enhances the clinical evaluation of patients with suspected major salivary gland neoplasms and will increase in availability. Salivary neoplasms are increasingly identified in imaging studies performed for other clinical indications. MRI will remain the imaging method of choice for most salivary gland neoplasms, with CT scan being complementary for some minor salivary gland tumors. Positron emission tomography scans can provide useful information regarding the extent of malignant tumors but have not been useful for primary tumor evaluation. It is expected that new functional imaging agents will be developed that will improve MRI utility. Other molecular imaging techniques using molecular probes will become available as new biomarkers are discovered.

The ability to histopathologically diagnose salivary gland neoplasms will also improve with the application of molecular methods to improve the sensitivity and specificity of imaging of biopsied or excised salivary gland tissue. The current trend of using ultrasound-guided core biopsies will require long-term assessment as to its safety, but its potential to cause tumor recurrence by tumor seeding is acceptable in the short term, before the technique becomes used universally.

The diagnosis of the recurrence of malignant salivary neoplasms will be aided by refinement of methods to detect serum biomarkers. This will play an important role in patient monitoring following definitive treatment and hopefully will allow for early detection of tumor recurrence, resulting in improved outcomes and prolonged patient survival for salvage surgery along with modern adjuvant treatments.

Minimally invasive surgical approaches will continue to evolve. New surgical instrumentation development will allow for refined surgical dissection with less tissue injury. Surgical robots are likely to become smaller and hence more applicable to usage in the upper aerodigestive tract and in skull base applications. Improved methods for cranial nerve monitoring are also likely to evolve. In addition, larger studies that are appropriately powered will be conducted to provide more data as to the utility of facial nerve monitoring in parotid surgery.

Improved methods of reconstruction continue to be developed. New free tissue transfer techniques that improve cosmetic and functional outcomes will result in improved outcomes and 'natural living' are evolving. There is renewed interest in the utility of using vascularized nerve grafts when cranial nerves are sacrificed or require rehabilitation, which may improve facial reanimation outcomes. Additionally, adjunctive methods for the correction of facial paralysis are being developed and will have a major impact on quality of life.

The delivery of radiation therapy will become more sophisticated, with improved tumor targeting based on enhanced treatment-planning methods. This will further reduce morbidity. In addition, the use of heavy particle therapy will continue to evolve for specific tumor applications.

Presently, chemotherapeutic agents and targeted therapies have so far been disappointing for the management of advanced primary high-grade, high-stage salivary cancers; recurrent locally advanced cancers; and metastatic local and distant malignant tumors. Through clinical trials, these agents' role as adjuvant therapies will become more obvious. It is hoped that new therapies that are more effective will become available. Their role in personalized cancer treatment is yet to be determined and is a large area for further investigation and research.

More quality-of-life studies are needed and will be reported, especially for patients with cancers of the salivary glands. The results of these studies will improve information delivery and will thus be important in the counseling of, selection of treatment for, and rehabilitation of patients with salivary gland neoplasms.

David W. Eisele, MD, FACS,
Baltimore, Md., USA
Patrick J. Bradley, MBA, FRCS,
Nottingham, UK

Author Index

Aygun, N. 25

Bradley, P.J. VII, 1, 9, 104, 113, 132, 175, 198
Byrne, P.J. 120

Chinn, S.B. 104
Clark, J. 95

Deschler, D.G. 83

Eisele, D.W. VII, 46, 83, 175, 198

Ferris, R.L. 113

Genther, D.J. 120
Gillespie, M.B. 53
Glastonbury, C.M. 25
Glazer, T.A. 182
Guntinas-Lichius, O. 46, 71, 120

Ha, P.K. 17
Howlett, D. 39

Iro, H. 53

Kiess, A.P. 157
Kontzialis, M. 25

Medina, J. 132
Mendenhall, W.M. 141
Merdad, M. 168

Nicolai, P. 63

Prestwich, R. 148

Quon, H. 157, 168

Richmon, J.D. 168

Sen, M. 148
Shuman, A.G. 182
Silver, N.L. 104
Slevin, N.J. 141
Stenman, G. 17

Talmi, Y.P. 189
Thomson, D.J. 141
Triantafyllou, A. 39

Vander Poorten, V. 71

Wang, S. 95
Wax, M.K. 189
Weber, R.S. 104
Witt, R.L. 63

Zbären, P. 132

Subject Index

Acinic cell carcinoma
    magnetic resonance imaging 27
    prognosis, see Prognosis, salivary gland carcinoma
ACC, see Adenoid cystic carcinoma
Adenocarcinoma (not otherwise specified)
    overview 14
    salivary gland distribution 11
    submandibular gland carcinoma 108
Adenoid cystic carcinoma (ACC)
    hormone receptor therapeutic targeting 154
    magnetic resonance imaging 34
    metastasis, see Metastasis
    molecular pathology 19, 20
    natural history and behavior 148, 149
    overview 13
    prognosis, see Prognosis, salivary gland carcinoma
    salivary gland distribution 11
    submandibular gland carcinoma 107, 108
ARID1A-PRKD1 gene fusion 18

Biopsy, see Fine needle aspiration cytology; Intra-operative frozen sections; Ultrasound-guided core biopsy
Bortezomib, salivary gland carcinoma management 154
Botulinum toxin, facial rehabilitation 128

CAMSG, see Cribriform adenocarcinoma of minor salivary glands
Carcinoma ex-pleomorphic adenoma (CXPA)
    imaging 32
    molecular pathology 21
    overview 13, 14
    salivary gland distribution 11
Cetuximab, salivary gland carcinoma management 153
CHCHD7-PLAG1 gene fusion 18
Chemotherapy
    combination therapy
        cisplatin/doxorubicin/5-fluorouracil 151
        cisplatin/mitoxantrone 152
        cyclophosphamide/cisplatin/pirarubicin 151
        cyclophosphamide/doxorubicin 152
        cyclophosphamide/doxorubicin/cisplatin 151
        cyclophosphamide/doxorubicin/cisplatin/5-fluorouracil 151
        gemcitabine/cisplatin 152
        overview 150, 151
        radiation therapy 146
    palliative care 186
    prospects 199
    recurrent salivary gland tumors 173
    single agents
        cisplatin 149, 150
        docetaxel 150
        epirubicin 150
        gemcitabine 150
        paclitaxel 150
        vinorelbine 150
Cisplatin, see Chemotherapy
Classification, salivary gland neoplasms
    histopathological grading classification 5
    historical perspective 2, 3
    histotyping classification 5
    mesenchymal tumors 3
    molecular biology and genetic classification 6, 7
    risk stratification 4, 5
    TNM staging 6, 7
    World Health Organization 2
Computed tomography (CT)
    inoperable malignancies 159
    overview 25
    perfusion studies 32
    perineural spread 27, 28
    posttreatment surveillance 28, 29
    recurrent salivary gland tumors 170
    submandibular gland carcinoma 105, 106

Cribriform adenocarcinoma of minor salivary glands (CAMSG), molecular pathology 22
CRTC1-MAML2 gene fusion 18, 19
CRTC3-MAML2 gene fusion 18
CT, *see* Computed tomography
CTNNB1-PLAG1 gene fusion 18
CXPA, *see* Carcinoma ex-pleomorphic adenoma
Cyclophosphamide, *see* Chemotherapy

DDX3X-PRKD1 gene fusion 18
Docetaxel, *see* Chemotherapy
Doxorubicin, *see* Chemotherapy

ED, *see* Extracapsular dissection
Electromyography, *see* Facial nerve monitoring; Facial reconstruction
Epidemiology, salivary gland neoplasms 1, 149
Epirubicin, *see* Chemotherapy
Epstein-Barr virus, salivary gland neoplasm risk factor 198
ETV6-NTRK3 gene fusion 18, 20
EWSR1-ATF1 gene fusion 18, 21
EWSR1-POUSF1 gene fusion 18, 19
Extracapsular dissection (ED), technique 59–61

Facial nerve monitoring
    benefits 49, 50
    electromyography 47, 48
    electrophysiology 47
    indications 48, 49
    limitations 50, 51
    litigation 51
    rationale 46, 47
    recurrent salivary gland tumor surgery 170
Facial reconstruction
    aims 128, 129
    facial function evaluation
        motor function 122
        nonmotor function 122, 123
    nonsurgical rehabilitation 128
    overview 120, 121
    palsy incidence 120
    preoperative assessment
        electromyography 121
        magnetic resonance imaging 122
    prospects 199
    recurrent salivary gland tumor surgery 171
    surgical rehabilitation
        eye closure restoration 128
        facial nerve suture and grafting 123–125, 194, 195
        free muscle transfer 126, 127
        muscle and tendon transposition 125, 126
        sling techniques 127, 128
        soft tissue defects 193, 194
    treatment planning 123, 124

FGF2, adenoid cystic carcinoma mutations 20
FGFR1-PLAG1 gene fusion 18
Fine needle aspiration cytology (FNAC)
    limitations 40, 41
    overview 40
    submandibular gland carcinoma 107
5-Fluorouracil, *see* Chemotherapy
FNAC, *see* Fine needle aspiration cytology
Frey's syndrome, salivary gland surgery patients 191, 192

Gefitinib, salivary gland carcinoma management 153
Gemcitabine, *see* Chemotherapy
Great auricular nerve, sacrifice in parotidectomy 190, 191

HCCC, *see* Hyalinizing clear cell carcinoma
HER2, salivary gland neoplasm overexpression 20, 21, 149
HGT, *see* High-grade transformation
High-grade transformation (HGT)
    overview 14
    salivary gland distribution 11
HMGA2-FHIT gene fusion 18
HMGA2-NFIB gene fusion 18
HMGA2-WIF1 gene fusion 18
Hyalinizing clear cell carcinoma (HCCC), molecular pathology 21

Imatinib, salivary gland carcinoma management 153
Intra-operative frozen sections (IOFSs), biopsy 40, 44
IOFSs, *see* Intra-operative frozen sections

Lapatinib, salivary gland carcinoma management 153
LIFR-PLAG1 gene fusion 18

Magnetic resonance imaging (MRI)
    benign neoplasms 33, 34
    diffusion-weighted imaging 30, 31
    dynamic contrast-enhanced imaging 31
    facial reconstruction 122
    high-resolution imaging 29, 30
    inoperable malignancies 159
    intravoxel incoherent motion imaging 31
    overview 25–27
    perineural spread 27, 28
    posttreatment surveillance 28, 29
    recurrent salivary gland tumors 170
    tumor staging 27
Magnetic resonance spectroscopy (MRS), salivary gland neoplasms 31
MALT lymphoma, imaging 34, 36
Mammary analog secretory carcinoma (MASC), molecular pathology 20
MASC, *see* Mammary analog secretory carcinoma

MEC, see Mucoepidermoid carcinoma
Melanoma metastasis, see Parotid gland neoplasm
Metastasis
    distant metastasis from salivary gland
        diagnosis 184, 185
        epidemiology 182, 183
        prognosis 184
        risk factors 183, 184
        sites 184
        treatment
            goals 185
            palliative care 186, 187
            supportive care 186, 187
    salivary gland neoplasm metastasis
        incidence 132
        management
            clinically negative neck 134–136
            clinically positive neck 133
            overview 149
        sentinel node biopsy 137, 138
        survival 132
    skin cancer to parotid gland, see Parotid gland neoplasm
Minor salivary gland neoplasm
    benign:malignant neoplasm ratio 10
    children 12
    distribution and management
        larynx 118
        nose 117, 118
        oral cavity 116
        oropharynx 116, 117
    imaging 54
    malignancy overview 114, 115
    pathology 54, 55
    pediatric surgery 179
    presentation 53, 54
    recurrent pleomorphic adenoma 69
    surgery
        benign tumors 55
        malignant tumors 115–117
    symptoms and signs 15
    types 11–13
Mitoxantrone, see Chemotherapy
MRI, see Magnetic resonance imaging
MRS, see Magnetic resonance spectroscopy
Mucoepidermoid carcinoma (MEC)
    imaging 35
    metastasis, see Metastasis
    molecular pathology 18, 19
    natural history and behavior 148, 149
    overview 2, 13
    pediatric 178
    prognosis, see Prognosis, salivary gland carcinoma
    salivary gland distribution 11

MYB-NFIB gene fusion 18–20

NOTCH, adenoid cystic carcinoma mutations 19, 20

PA, see Pleomorphic adenoma
Paclitaxel, see Chemotherapy
Parotid gland neoplasm
    benign:malignant neoplasm ratio 10
    carcinoma prognosis, see Prognosis, salivary gland carcinoma
    children 12
    follow-up 169
    imaging 58, 59
    intraparotid lymph node imaging 36
    malignant tumors
        epidemiology 83, 84
        evaluation 84, 85
        presentation 84
    metastasis, see also Metastasis
        melanoma
            adjuvant therapy 101
            epidemiology 98
            neck management 100
            prediction 98–100
            surgery 100
        sentinel node biopsy 137, 138
        squamous cell carcinoma
            epidemiology 95, 96
            prediction 96
            prognosis 96, 97
            surgery 97, 98
    pathology 59
    pediatric surgery 179
    presentation 58
    surgery
        benign tumors
            extracapsular dissection 59–61
            partial parotidectomy 59
            superficial parotidectomy 59
        malignant tumors
            facial nerve management 90
            neck management 90, 91
            outcomes 91
            technique 85–89
    surgical incision and quality of life outcomes 192
    symptoms and signs 14, 15
    types 11–13
Pediatric salivary gland neoplasms
    differential diagnosis 176, 177
    evaluation 177
    incidence 175
    malignant tumors 178
    outcomes 180
    pleomorphic adenoma 178

presentation 177
recurrence 179, 180
staging 178
surgery
    neck dissection 179
    parotid gland 179
    sublingual and minor salivary glands 179
    submandibular gland 179
tissue distribution 176
Perineural spread (PNS)
    imaging 27, 28
    inoperable malignancies 158
    prognostic value 75
PET, see Positron emission tomography
PIK3CA, salivary duct carcinoma mutations 20
Pirarubucin, see Chemotherapy
Pleomorphic adenoma (PA)
    computed tomography 33
    magnetic resonance imaging 26
    molecular pathology 21
    pediatric 178
    recurrent adenoma
        distribution 69
        etiology 64, 65
        presentation 63
        radiation therapy 67, 68, 142, 143
        surgery 65, 67
    salivary gland distribution 11, 12
PLGA, see Polymorphous low-grade adenocarcinoma
PNS, see Perineural spread
Polymorphous low-grade adenocarcinoma (PLGA)
    molecular pathology 22
    overview 13
    salivary gland distribution 11
Positron emission tomography (PET)
    salivary gland neoplasm imaging 28, 30, 32, 33
    submandibular gland carcinoma 107
PRKD1, activating mutations 22
Prognosis, salivary gland carcinoma
    clinical translation of prognostic scores 80
    limitations of prognostic index system 80, 81
    nomogram creation 76, 78
    oncological outcomes and specific prognostic factors 72
    overview 71, 72
    parotid carcinoma
        prognostic index development 74–76
        survival analysis 72, 73
    perineural growth as prognostic factor 75
    prospects 81
    research echelons 72
    summary index creation 73, 74

validation of prognostic scores
    clinical validation 78–80
    external validation 78
    statistical validation 80

QoL, see Quality of life
Quality of life (QoL)
    overview 189, 190
    salivary gland surgery patients
        cosmesis 193, 194
        facial paralysis 194, 195
        Frey's syndrome 191, 192
        incisions and outcomes 192, 193
        salivary gland carcinoma patients 195, 196
        sensation loss 190, 191

Radiation therapy
    adjuvant indications 171, 172
    indications
        benign disease 141–143
        malignant disease
            charged particle therapy 145, 146
            chemotherapy combination 146
            late toxicities of parotid bed radiotherapy 144, 145
            postoperative 143, 144
            primary radiotherapy 145
    inoperable malignancy management
        carbon ion radiation 162
        chemoradiation 163, 164
        follow-up 165
        hyperfractionated photon radiation 163
        hypofractionated stereotactic photon radiation 163
        neutron radiation 160, 161
        overview 172
        palliation 164, 186, 187
        photon radiation 159, 160
        proton radiation 162, 163
        stereotactic radiosurgery 61, 162
        techniques 164, 165
    prospects 199
    recurrent pleomorphic adenoma 67, 68
    recurrent salivary gland tumors 172, 173
    salivary gland neoplasm metastasis
        clinically negative neck 134–136
        clinically positive neck 133

Salivary duct carcinoma (SDC)
    hormone receptor therapeutic targeting 154
    magnetic resonance imaging 28
    metastasis, see Metastasis
    molecular pathology 20
    natural history and behavior 148, 149

overview 14
salivary gland distribution 11
SDC, *see* Salivary duct carcinoma
Sorafenib, salivary gland carcinoma management 154
SPEN, adenoid cystic carcinoma mutations 20
Squamous cell carcinoma metastasis, *see* Parotid gland neoplasm
Sublingual gland neoplasm
    anatomy 113, 114
    benign:malignant neoplasm ratio 10
    children 12
    pathology of malignant tumors 114
    pediatric surgery 179
    presentation 55, 56
    prognosis 114
    surgery
        benign tumors 56, 57
        malignant tumors 114
    symptoms and signs 15
    types 11–13
Submandibular gland neoplasm
    benign:malignant neoplasm ratio 10
    children 12
    fine needle aspiration biopsy 107
    histopathology of malignant tumors 107, 108
    imaging 57, 105–107
    pediatric surgery 179
    presentation 57
    recurrent pleomorphic adenoma 69
    surgery
        benign tumors 57, 58
        incision and quality of life outcomes 192
        malignant tumors 108–110
        neck management 111
    symptoms and signs 15, 105
    types 11–13
Sunitinib, salivary gland carcinoma management 154

Tamoxifen, salivary gland carcinoma management 154
TCEA1-PLAG1 gene fusion 18
TNM staging
    imaging 27
    pediatric malignancies 178
    salivary gland neoplasms 6, 7
Trastuzumab, salivary gland carcinoma management 154

UGCB, *see* Ultrasound-guided core biopsy
Ultrasound
    overview 25
    pediatric salivary gland neoplasms 177
Ultrasound-guided core biopsy (UGCB)
    advantages 42
    controversies 42–44
    overview 41, 42

Vinorelbine, *see* Chemotherapy

Warthin tumor
    imaging 30, 33
    overview 12
    recurrence 69
    salivary gland distribution 11
WHO, *see* World Health Organization
World Health Organization (WHO), salivary gland neoplasm classification
    overview 2
    risk stratification 4, 5